W9-BLM-541

Sustainable Communities and the Challenge of Environmental Justice

Sustainable Communities and the Challenge of Environmental Justice

Julian Agyeman

NEW YORK UNIVERSITY PRESS
New York and London

NEW YORK UNIVERSITY PRESS
New York and London
www.nyupress.org

Library of Congress Cataloging-in-Publication Data
Agyeman, Julian.
Sustainable communities and the challenge of environmental justice /
Julian Agyeman.
p. cm.
Includes bibliographical references and index.
ISBN 0-8147-0710-6 (cloth : acid-free paper)
ISBN 0-8147-0711-4 (pbk. : acid-free paper)
1. Environmental justice. 2. Sustainable development. I. Title.
GE220.A34 2005
363.7—dc22 2005003742

New York University Press books are printed on acid-free paper,
and their binding materials are chosen for strength and durability.

Manufactured in the United States of America

c 10 9 8 7 6 5 4 3 2 1
p 10 9 8 7 6 5 4 3 2 1

For Janet Clayton and women worldwide fighting breast cancer.

Contents

Acknowledgments

It may sound like a cliché, but this book was written by many people, not just me. It is the product of hundreds of discussions, arguments, and readings; personal and collective experiences; three academic papers; and far too many late nights. However, in this endeavor, I have numerous people to thank.

First, Rachel Bratt, who, as chair had the confidence to hire me to the Tufts Department of Urban and Environmental Policy and Planning (UEP) in 1999. It is a privilege to be a part of such a scholarly community, where I have been able to develop my ideas in a very supportive environment. Not only did she hire me, but she gave me $5,000 for my first research assistant, UEP student Tom Evans, who contributed greatly to the paper that eventually became chapters 3 and 4.

A few years later, in 2001, Tufts University College of Citizenship and Community Service, through the leadership of Rob Hollister and Molly Mead, great supporters of my work, gave me a $10,000 grant to hire a research assistant, UEP student Briony Angus, who contributed her ideas and research to an article on civic environmentalism that became part of chapter 2. In 2004, the Tufts Faculty Research Awards Committee, under the guidance of chair Andrew McLellan gave me $1,500 toward hiring a research assistant, UEP student Jonathan Grosshans, who so ably helped me carry out the research for my case study of Alternatives for Community and Environment in chapter 5. Thanks also go to UEP student Kirstin Henninger for her invaluable research on deliberative democracy and for her wonderful Austin Powers impressions, and to Jim Coburn for editing my manuscript.

Thanks also go to other colleagues at Tufts. To deans Susan Ernst and Kevin Dunn for agreeing to Junior Faculty Research Leave in the spring semester of 2004; to Sheldon Krimsky for his support, guidance, and invaluable advice; to Francine Jacobs for her candor and sense of humor; to James Jennings for his insights into Roxbury; to Veronica

Eady, now of West Harlem Environmental Action, for immersing me, an outsider in the U.S. environmental justice scene, with such grace and ease; to Kent Portney for his insights into sustainable communities; and to Dale Bryan for both involving me in the Mystic Watershed Collaborative and contributing to chapter 1.

I am indebted to Bob Evans (Northumbria University, UK) and Robert D. Bullard (Clark Atlanta University), co-authors of a paper that influenced chapters 3 and 4, and to the group of critical readers I asked to critique early drafts of my (incomplete) manuscript. Kee Warner (University of Colorado, Colorado Springs), David Pellow (University of California, San Diego), Robert Brulle (Drexel University), Sheila Foster (Fordham University), David Schlossberg (Northern Arizona University), J. Timmons Roberts (College of William and Mary), JoAnn Carmin (MIT), Jennifer Hammer (NYU Press), Veronica Eady, and Kent Portney, your comments were perceptive and well made, and any inaccuracies or advice I did not follow are my fault, and my loss.

During spring and early summer 2004, it was my privilege to work with a group of people who I admire greatly, the staff and board of ACE, who gave up their time so freely for me: Bill Shutkin and Charlie Lord (founders), Bob Terrell, James Hoyte, Lisa Goodheart, Gary Gill-Austern, Penn Loh, Warren Goldstein-Gelb, Jodi Sugerman-Brozan, Quita Sullivan, Eugene Benson, Khalida Smalls, and Klare Allen. Marlena Rose, Celina Lee, and Alma Feliciano, forgive me for not interviewing you, but I simply ran out of time. Keep on changing the world, guys.

Finally, to the person without whose support and love I could never have freed myself up to get an MA, let alone a PhD and (hopefully) tenure, my wife, Lynn Graham.

Introduction

> ACE as an environmental justice group has always struggled with its relationship with the more traditional or mainstream environmental and sustainability groups. We've played with them. "Clean Buses for Boston" was quite intentionally, on my part, an effort to reach out to more mainstream groups in coalition, to bond with them. . . . Frankly, we didn't need them, but we were doing similar work. . . . The Boston Foundation stepped up and then the Public Welfare Foundation and everybody stepped up because what were we doing? We were bringing neighborhood environmental justice organizations together with mainstream environmental and sustainability organizations.
>
> —Bill Shutkin, co-founder,
> Alternatives for Community and Environment

The relationship between environmental justice and sustainability groups has traditionally been uneasy. What might at first glance seem like an obvious case for partnership, for coalition, is fraught with ideological and other concerns, despite the obvious enthusiasm of funders. How has it come to this, and more to the point, how do we move forward?

Environmental Justice and Sustainability

Environmental justice and sustainability are two concepts that have evolved over the past two decades to provide new, exciting, and challenging directions for public policy and planning. *Environmental justice* can be understood as a local, grassroots, or "bottom-up" community reaction to external threats to the health of the community, which have been shown to disproportionately affect people of color and low-income neighborhoods. *Sustainability,* on the other hand, refers to meeting our needs today while not compromising the ability of those that follow to meet their needs. It emerged in large part from "top-down" international processes and committees, governmental structures, think tanks, and international nongovernmental organization (NGO) networks, although it is now, like environmental justice, at the grassroots level that much-needed change is happening. Both concepts are highly contested and problematized, but they nevertheless have tremendous potential to effect long-lasting change on a variety of levels, from local to global.

Environmental justice organizations emerged from grassroots activism in the civil rights movement. Whether these organizations are based on neighborhood, community, university, or region and whether they are staffed or unstaffed, they have expanded the dominant traditional[1] environmental discourse, based around environmental stewardship, to include social justice and equity considerations. In doing this, they have redefined the term *environment* so that the dominant wilderness, greening, and natural resource focus now includes urban disinvestment, racism, homes, jobs, neighborhoods, and communities. The *environment* became discursively different; it became "where we live, where we work and where we play" (Alston 1991). The environmental justice movement has been, and continues to be, very effective at addressing the issues of poor people and people of color, who are disproportionately affected by environmental "bads" such as toxic facilities, poor transit, and increased air pollution and who have restricted access to environmental "goods" such as quality green and play spaces.

At the same time, sustainable development and sustainable community advocates have mostly, but not exclusively, come from the traditional environmental movement[2] and are generally professionally qualified, often in a cognate discipline. They are usually from a different social location from people in the environmental justice movement. Wary of interest-group pluralism, where individuals in groups become the principal actors in democratic politics, with its attendant problem of capture, or domination, by powerful interests, sustainability advocates promote the use of innovative deliberative and democratic processes. These so-called deliberative and inclusionary processes and procedures (DIPS) are being increasingly used in Europe, North America, and, more recently, Australia.

DIPS include visioning, study circles, collaboration, consensus building and consensus conferencing, negotiation and conflict resolution, and citizen's juries. The overall aim is to involve a broad cross-section of lay citizens in the development of shared values, consensus, and a vision of the common good. This deliberative focus is integral to the sustainable development and sustainable communities project (Renn et al. 1995; Dryzek 1990; Smith 2003). As a very general rule, DIPS differentiate sustainable development and sustainable communities organizations from much of the environmental justice movement in that sustainable community advocates tend, through deliberation, to be more proactive in saying what kind of communities we should be aiming for. Most but not all groups in the environmental justice movement are trapped in the traditional pluralistic decision-making processes, common in much environmental law, that make reaction the norm and proaction much more difficult.

Indeed, much of the activity of the environmental justice movement, certainly the small neighborhood groups as distinct from the movement's professional[3] not-for-profits and university centers, is *reactive*—that is, focused on stopping environmental bads as they threaten the community. This is not what the Principles of Environmental Justice (see Appendix), the theoretical and ideological foundation of the movement advocate, but reaction is the political reality for many communities starved of resources. The purveyors of environmental bads, such as large multinationals, are favored in pluralistic decision-making processes because of their disproportionate influence, economic muscle, and knowledge. This David-and-Goliath struggle has nevertheless propelled the movement a long way over the past twenty years. Where the movement has been less successful, though not completely unsuccessful, is in developing consensual visions and taking ownership of the assets and resources necessary to bring such visions to fruition.

Cooperation?

Despite the historical and geographical differences in origin between *environmental justice* and *sustainability*, there exists an area of theoretical, conceptual, and practical compatibility between them. Each concept has its own particular discursive frame[4] and paradigm,[5] which can be seen as opposite ends of a continuum. At one end is the Environmental Justice Paradigm (EJP) of Taylor (2000), which is a framework for integrating class, race, gender, environment, and social justice concerns.

It represents the theoretical underpinning of the environmental justice project and activism. At the other end is the New Environmental Paradigm (NEP) of Catton and Dunlap (1978), which sets out an environmental stewardship and sustainability agenda that currently influences the work of most environmental and sustainability organizations but has little to say about equity or justice. Notwithstanding these differences, which are primarily about the issues of race and class, justice and equity, and not about the need for greater environmental protection, there is a rich and critical nexus where proponents of each paradigm are engaging in *cooperative endeavors* (Schlosberg 1999) regarding common issues such as toxics use reduction and transportation.

Yet such cooperation has so far largely been based on what Gould et al. (2004:90) call "short-term marriages of convenience" rather than "longer-term coalitions." In this respect the cooperation currently falls well short of Cole and Foster's (2001:164) concept of *movement fusion,* "the coming together of two (or more) social movements in a way that expands the base of support for both movements by developing a common agenda." If and when this happens, the result may be a broad, integrated social movement to create just and sustainable communities for all people in the future. This possibility is the inspiration for this book.

In order for the environmental justice and sustainability movements to develop a common agenda, changes to both will be required. One change is already happening within the sustainability paradigm, in part as a result of the influence of the environmental justice project. It is the emergence of a "just sustainability" orientation as a counter to the dominance of "environmental sustainability." This development is the focus of this book.

Just Sustainability

In the fall of 2002, the eleven-year anniversary of the landmark 1991 National People of Color Environmental Leadership Summit took place in Washington, D.C. Earlier that year, the World Summit on Sustainable Development (WSSD), the ten-year follow-up to the 1992 United Nations Conference on Environment and Development (UNCED), took place in Johannesburg, South Africa. Despite battles in Johannesburg between the "green" or *environmental* agenda of wealthy countries representing the North and the "brown" or *antipoverty* agenda of poorer countries representing the South, the discourse of both conferences revolved around what Agyeman et al. (2003) have termed *just sustainability,* or what Jacobs (1999:32) calls "the egalitarian conception of sustainable development." The concept of *just sustainability* highlights the pivotal role that justice and equity could and should play within sustainability discourses. In so doing, it fundamentally challenges the current, dominant, stewardship-focused orientation of sustainability, which has as its main concern the conservation of the natural environment, namely *environmental* sustainability (Dobson 1999, 2003), or what Jacobs (1999:33) calls "the non-egalitarian conception" of sustainable development.

Why should race and class, justice and equity play a role in sustainability? Has the environmental sustainability movement not done a good job? No, according to Shellenberger and Nordhaus (2004:12) in *The Death of Environmentalism*:

> Why, for instance, is a human-made phenomenon like global warming —which may kill hundreds of millions of human beings over the next century—considered "environmental"? Why are poverty and war not considered environmental problems while global warming is? What are the implications of framing global warming as an environmental problem—and handing off the responsibility for dealing with it to "environmentalists"?

Irrespective of whether we take a global, statewide, or more local focus, a moral or practical approach, inequity and injustice resulting from, among other things, racism and classism are bad for the environment and bad for sustainability. What is more, the environmental sustainability movement, typified by the National Audubon Society, the World Wildlife Fund (WWF), and the Natural Resources Defense Council, does not have an analysis or theory of change with strategies for dealing with these issues. While researching a film in the early 1990s, I asked a Greenpeace UK staffer if she felt that her organization's employees reflected multicultural Britain. She replied calmly, "No, but it's not an issue for us. We're here to save the world."

Yet research has shown that, globally, nations with a greater commitment to equity and a correspondingly more equitable society tend to also have a greater commitment to environmental quality (Torras and Boyce 1998). Good examples are the Nordic countries of Sweden,

Denmark, Norway, and Finland. In a survey of the fifty U.S. states, Boyce et al. (1999) found that those with greater inequalities in power distribution (measured by voter participation, tax fairness, Medicaid access, and educational attainment levels) had less stringent environmental policies, greater levels of environmental stress, and higher rates of infant mortality and premature death. At a more local level, a study by Morello-Frosch (1997) of counties in California showed that highly segregated counties, in terms of income, class, and race, had higher levels of hazardous air pollutants.

If sustainability is to become a process with the power to transform, as opposed to its current environmental, stewardship, or reform focus, justice and equity issues need to be incorporated into its very core. This, as the title of my book suggests, is the gauntlet the environmental justice movement has thrown down to the development of sustainable communities. Our present green or environmental orientation of sustainability is basically about tweaking our existing policies. Transformative or just sustainability implies a paradigm shift that requires sustainability to take on a redistributive function. To do this, justice and equity must move center stage in sustainability discourses, if we are to have any chance of a more sustainable future.

The Just Sustainability Paradigm (JSP) is an emerging discursive frame and paradigm. It is not, however, rigid, single, and universal; it is linked to both the EJP and the NEP. In this sense, the JSP can be seen as both flexible and contingent, composed of overlapping discourses that come from recognition of the validity of a variety of issues, problems, and framings. The JSP arises from the definition of sustainability of Agyeman et al. (2003:5), "the need to ensure a better quality of life for all, now and into the future, in a just and equitable manner, whilst living within the limits of supporting ecosystems," a definition that prioritizes justice and equity but does not downplay the environment, our life-support system. In essence, the JSP is malleable, acting as a bridge spanning the continuum between the EJP and the NEP.

While it is growing in acceptance, just sustainability has not yet been recognized as a pivotal concept by all scholars in the field. In 2000, Brulle created a typology of nine different discursive frames within the U.S. environmental movement: manifest destiny, wildlife management, conservation, preservation, reform environmentalism, deep ecology, environmental justice, eco-feminism and eco-theology. He did not identify just sustainability as a frame, nor did he name sustainable development as one. Instead he posited sustainable development as a subset of the conservation frame, with the note that "although sustainable development may be a latter day version of conservation, it has interjected ecological concerns into industry" (Brulle 2000:158).

However, I believe that sustainability as the theoretical component and sustainable development and sustainable communities as the practical components is a far more durable, influential frame than Brulle's research shows (cf. Campbell 1996), especially outside the United States. Indeed, what seems remarkable is that nearly twenty years on, the Brundtland Report, which popularized the term *sustainable development* (World Commission on Environment and Development 1987), and agreements made at UNCED in 1992 are still having cross-sectoral influence. And this influence is increasing. For example, greater numbers of multinationals (such as BP, now tellingly "Beyond Petroleum," and IKEA), national governments (the Netherlands and Denmark plan to be sustainable by 2050), and local governments (250 North American cities, out of 600 globally, have signed up to the Cities for Climate Protection Plan) seem to be taking sustainability increasingly seriously.

While most are firmly in the NEP, the JSP influences the work of a few national environmental and sustainability membership organizations, and there is a growing number of local organizations, programs, and projects that utilize the discursive frame and paradigm of just sustainability to practical effect in U.S. cities, as I shall demonstrate. This paradigm underpins the leading-edge cooperative endeavors that are described in this book. Further down the line, it will be both a *precondition* for movement fusion between the environmental justice and sustainability movements and the *cement* that keeps the coalition together.

The emergent JSP is a far bigger tent than could be filled solely by just sustainability and by most environmental justice organizations. My work concentrates on this paradigm and the movements and organizations that espouse it because they are founder members. I hope that future researchers who want to go further in characterizing the JSP will look more broadly toward initiatives such as the Just Transition Alliance, "a voluntary coalition of labor, economic and environmental justice activists, Indigenous people and working-class people of color [that] has created a dialogue in local, national, and international arenas."[6] This and many other alliances that are forming around the world could, I believe, unite under the JSP.

In order to be truly successful at both alleviating environmental bads and bringing about substantial community-envisioned change, the environmental justice movement and its organizations will have to take a more proactive, deliberative, sustainable communities–type visioning approach. While support for this approach is now gaining traction, the antecedents of this thinking go back at least as far as 1993, when Goldman (1993:27) suggested that "sustainable development may well be seen as the next phase of the environmental justice movement."

At the same time, the sustainable development and sustainable community movement will need to fully respond to the environmental justice movement's ongoing critique of its overeagerness to focus on *environmental* sustainability rather than on a more holistic conception of sustainability that sees justice and equity, and the interlinkages between environmental, economic, and social issues, as the necessary focus of activism. When these conditions are met, as I believe they are beginning to be in organizations using the JSP as their frame, Cole and Foster's (2001) concept of movement fusion has a chance of taking place. Indeed, Cole and Foster see such fusion as "a necessary ingredient for the long term success of the environmental justice movement because, put simply, environmental justice advocates do not have a large enough power base to win the larger struggle for justice on their own" (165).

Coalition building through a convenient common focus such as toxics, transit, or antinuclear issues is not without precedent. Antiglobalization protests in Seattle, London, and Washington, D.C., among others, consisted of a range of environmental, peace, indigenous, spiritual, women's, civil rights, labor, antiracist, and other groups who would also be candidates for the JSP. Gould et al. (2004) have begun to investigate what they call "Blue-Green" or "Seattle" coalitions between labor unions and environmentalists. To do this they pose four fundamental problems, the last three of which we will be returning to in later chapters in this book: "(1) the problem of reciprocation and unbalanced expectations by environmentalists for unionists; (2) the problem of extending short term marriages of convenience into longer term coalitions; (3) the debate over whether local or national levels are better places to make these coalitions; and (4) the class issue" (96). The researchers' conclusion is that, because of fewer ideological obstacles and the structural positions and origins of such groups, "the environmental justice and environmental health wings of the green movement are more suited to making long term coalitions with labor than are habitat-oriented green groups" and that "in many ways, the tensions between labor and the mainstream greens echo the tensions between the environmental justice movement and mainstream greens" (108). This reasoning is precisely why, as I will show, the best chance for more cooperative endeavors and ultimately movement fusion between the environmental justice and sustainability movements will be for environmental justice groups to work with just sustainability groups, as opposed to those of an environmental sustainability orientation.

There is good news and bad news on the environmental justice and just sustainability coalition front. The good news is that there is evidence that the JSP, which links the frames, concepts, language, programs, and repertoire of action[7] of organizations in the environmental justice and sustainability movements, is already emerging at the local, national, and international levels. This linkage is happening more within and through larger environmental-justice-based organizations such as Alternatives for Community and Environment (ACE) in Boston, Center for Neighborhood Technologies (CNT) in Chicago, and Urban Ecology in Oakland, California, than in smaller, neighborhood-type environmental justice groups which often do not have the time or the resources. ACE, CNT, Urban Ecology, and other leading-edge organizations of the JSP are being both reactive and proactive: they are operating within an environmental justice framework[8] (Bullard 1994) but are also exploring the wider and emerging terrain of sustainable development and the development of sustainable communities. At the national level, membership-based groups such as the Center for Health, Environment and Justice, Environmental Defense, Center for a New American Dream, and Redefining Progress, and internationally, the Earth Council, the Heinrich Boll Foundation, and the Stockholm Environment Institute, among others, are espousing the language, framing, and paradigm of just sustainability.

The bad news is that local governments, which were charged at UNCED in Rio in 1992 with delivering Local Agenda 21,[9] a community-led plan for local sustainability, are not making as much progress as local, national, or international NGOs are in this linkage.[10] In a study of sustainability projects in the largest U.S. cities, Warner (2002) found that few even acknowledged environmental justice as an aspect of sustainability. Similarly, the Environmental Law Institute (1999) analyzed 579 applications to the Environmental Protection Agency's (EPA) 1996 Sustainable Development Challenge Grant Program. Fewer than 5 percent of applications had "equity" as a goal, and interestingly, fewer than 1 percent addressed "international responsibility" through local-global linkages.

Theory, Method, and Analysis

This book is meant for people in a range of academic disciplines in the social and political sciences, the environmental sciences, environmental justice, environmental policy and planning, geography, and sustainability and for readers who do not identify themselves as part of any discipline, be they practitioners, activists, or the like.

To characterize and chart the rise of the JSP with a theoretical, methodological, and analytical rigor and robustness that is both acceptable and understandable to a diverse audience and at the same time useful to them is no easy task. The literatures drawn on in this study are wide-ranging. In essence, the book takes both a discourse analytic and interpretive approach to the emergence of the JSP, fully characterizing it and differentiating it from the discursive frames of both the NEP and EJP. One could argue that discourse, in and of itself, is no basis to make such claims. Brulle (2000:97), however, argues that "the discourse of a movement translates the historical conditions and the potential for mobilization into a reality that frames an organization's identity. This identity then influences the organization's structure, tactics and methods of resource mobilization." Carmin and Balser (2002:371) add another, interpretive approach in that "experience, core values and beliefs, environmental philosophy, and political ideology contribute to interpretive processes that take place within Environmental Movement Organizations (EMOs) that in turn shape the selection of a repertoire." In essence these contributors act as filters that affect how the political environment is interpreted by an organization, the programs it develops, and the actions it takes. Put another way, "these filters lead to interpretations and the construction of meaning that in turn can provide a foundation for action" (Carmin and Balser 2002:367). My research uses a combination of discourse analysis and an interpretive approach that I believe will give a clearer picture.

I use a content analytic approach based on available literature to differentiate the discourse of civic environmentalism (the dominant subnational environmental policymaking discourse) into two foci: narrow and broad. Narrow-focus civic environmentalism represents the NEP:

business as usual, reform, or unreconstructed (Agyeman 2001) environmentalism. Broad-focus civic environmentalism represents a more politically based construct, namely that environmental quality and economic and social health are inextricably interlinked (Shutkin 2000). In this sense, in discourse and issue framing, it is close to the transformative JSP.

In order to assess organizations' commitment to the JSP, I have created a Just Sustainability Index (JSI). I used a hybrid of discourse analysis, content/relational analysis, and interpretive analysis. The JSI assesses the discourse of organizations through the language and meanings inherent in their mission statements and in prominent contemporary textual or programmatic material available on the Internet. In my experience on the boards of several environmental and sustainability organizations in the United States and Europe, mission statements have been like meditational mantras, from which organizational and individual action and work plans flow. They could, however, be criticized as a form of purely aspirational discourse, rather than being based on "experience, core values and beliefs, environmental philosophy, and political ideology" (Carmin and Balser 2002:371). For this reason, I also looked at prominent contemporary textual or programmatic material that specifies the programs an organization will implement or has implemented and the resulting actions it will take or has taken in pursuance of program goals. This material represents an organizational interpretation of the political environment, and what the organization intends to do about it.

Based on the inclusion or exclusion of certain key words or concepts in the organizational mission and programs, and their relation to other key words or concepts, the JSI assigns organizations a score on a scale of 0 to 3, where 0 means that there is no mention of equity or justice in the core mission statement or in prominent contemporary textual or programmatic material and 3 means that the core mission statement relates to intra- and intergenerational equity and justice, and/or justice and equity occur in the same sentence in prominent contemporary textual or programmatic material.

I analyzed thirty top national environmental and sustainability membership organizations in this way. My intent was solely to provide support for my assertions about the dominance in the United States of the stewardship-focused orientation of sustainability, or of environmental sustainability over just sustainability. In addition, I selected three vignettes in each of the following sustainability categories: land-use planning, solid waste, toxic chemical use, residential energy use, and transportation. These vignettes are of representative programs or projects, managed by organizations with a JSI of 3, that are providing proactive, balanced efforts to create a just sustainability in practice in U.S. cities. I therefore include in this book a total of fifteen vignettes, each of which illustrates different practical aspects of the JSP.

In order to move beyond discourse and interpretation—from words to deeds—I also employ a case-study approach to provide a rich description of one organization and its programs that I believe in many ways represent the JSP. ACE in Boston has been attempting to create change, initially in the city's Roxbury district but latterly on a more regional basis. I assess the organization's links to the discourse, framing, and paradigm of just sustainability through a variety of sources of evidence, such as programs, documents, archival records, participant observation at meetings, and interviews with staff and board members.

The goal of this book is twofold:

- To characterize and illustrate the discourse of the JSP. I will illuminate the nexus between the concepts of sustainability and environmental justice both theoretically and by presenting a range of local or regionally based practical urban models in land-use planning, transportation, residential energy use, solid waste, and toxic chemical use.
- To identify an organization engaging with the JSP. Boston's ACE works locally, within an environmental justice framework, but is increasingly taking a more proactive, (metro) regional, systemic sustainable communities–type approach in creating alternative visions and solutions. I explore its programs and repertoires, including tools, techniques, and strategies, through an in-depth case study;

Chapter 1 takes a brief tour through the historical construction and discourse(s) of environmental justice in the United States. It looks at the Principles of Environmental Justice both as the source of inspiration and unison in the movement and also as the site of a major cleavage between activists and academics. The chapter continues by defining and framing environmental justice and looking at the EJP. It concludes by looking at the issues inherent in developing environmental justice policy.

Chapter 2 provides an overview of the history of sustainable development as currently practiced through the NEP. It problematizes this current practice through an examination of the pivotal role of justice and equity, and of "new economics," then moves on to look at the characteristics of a sustainable community and the discourse(s) of civic environmentalism and their relevance to sustainable communities.

Chapter 3 looks at the differences between the discourses of the JSP, the EJP, and the NEP. It concludes that there are five key differences between the JSP and the EJP: the JSP has a central premise of developing sustainable communities; the JSP has a wider range of progressive, proactive, policy-based solutions and policy tools; the JSP is calling for, and has developed, a coherent "new economics"; the JSP has much more of a local-global linkage; and the JSP is more proactive and visionary than the typically reactive EJP.

Chapter 4 develops the Just Sustainability Index, which can be used to assess an organization's discourse and texts and thereby its stated commitment to the JSP. The chapter continues by investigating the JSP through practical urban examples in the issue areas of land-use planning, transportation, residential energy use, solid waste, and toxic chemical use.

Chapter 5 offers an in-depth case study of Boston's ACE in order to provide a rich description of one organization that, while historically working within an environmental justice framework, is actively exploring the JSP.

Chapter 6 asks where we are now and if we have a map of where we need to go to develop more long-term cooperative ventures and ultimately movement fusion.

1 Environmental Justice

In this chapter, I attempt to do three things that are necessary in order for the reader to understand my later arguments and case study. First, I briefly track the history of the environmental justice (EJ) concept and resulting movement. I examine its institutional setting and some of the policy tools its advocates and activists use, finally offering an EJ critique of risk assessment and expert-led research. Second, I look at the definition, framing, and discourse of environmental justice and at the EJP in order to compare and contrast it to the NEP and JSP in chapter 3. Third, I look at unequal environmental protection in Massachusetts, as this is the physical setting for my case study of ACE in Boston in chapter 5, focusing on metro Boston's Mystic River Watershed, the development and implementation of the Commonwealth's policy, the difficulties inherent in defining "Environmental Justice Populations," and the lack of explicit policy linkages to the state's sustainable development policies.

A Brief History

Environmental justice *concerns* have been around since the conquest of Columbus in 1492.[1] The U.S. environmental justice *movement,* however, is generally believed to have started around fall 1982, when a large protest took place in Warren County, North Carolina.[2] The state wanted to dump more than six thousand truckloads of soil contaminated with PCBs into what was euphemistically described as "a secure landfill." The protesters came from miles around. They were black and white, ordinary (outraged) citizens, and prominent members of the civil rights movement and the National Black Caucus. Police arrested more than five hundred protestors in what Geiser and Waneck (1994:52) describe as "the first time people have gone to jail trying to stop a toxic wastes landfill."

As a result of the events in Warren County, the General Accounting Office (GAO) examined the location of four hazardous-waste landfills in EPA Region IV (the Southeast), where racial minorities average 20 percent of the total population. However, the four facilities were found to be in communities in which minorities made up 38 percent, 52 percent, 66 percent, and 90 percent of the population. The GAO concluded that there was enough evidence to be concerned about inequities in facility siting (GAO 1983). The landmark 1987 United Church of Christ study "Toxic Wastes and Race in the United States" showed that certain communities, predominantly communities of color, are at disproportionate risk from commercial toxic waste. This finding was confirmed by later research (Adeola 1994; Bryant and Mohai 1992; Bullard 1990a, 1990b; Mohai and Bryant 1992; Goldman 1993). It also led to the coining of a term, by Benjamin Chavis, that became a rallying cry: *environmental racism.*

The finding of environmental racism combined with the conclusion of Lavelle and Coyle (1992) in the *National Law Journal* that there is unequal protection and enforcement of environmental law by the EPA, has ensured that there is now a full-fledged environmental justice movement made up of tenants associations, religious groups, civil rights groups, farm workers, professional not-for-profits, university centers and academics, and labor unions, among others. The movement stretches from Alaska to Alabama and from California to Connecticut, driven by the grassroots activism of African American, Latino, Asian and Pacific American, Native American, and poor white communities. As such, according to Pulido (1996a), it is a multiracial movement which is organizing around locally unwanted land uses (LULUs) such as waste facility siting, transfer storage and disposal facilities (TSDFs), and other issues such as lead contamination, pesticides, water and air pollution, workplace safety, and transportation. More recently, issues such as sprawl and smart growth (Bullard et al. 2000), sustainability (Agyeman et al. 2003), and climate change (International Climate Justice Network 2002; Congressional Black Caucus Foundation 2004) have become targets for the environmental justice critique.

The movement's base, like that of the sustainability movement, has many foundations. According to Cole and Foster (2001), these are the civil rights movement, the antitoxics movement, academia, Native American struggles, the labor movement, and the traditional environmental movement.[3] They call the environmental justice movement "a movement based on environmental issues but situated within the history of movements for social justice" (31). Faber (1998:1) calls it "a new wave of grassroots environmentalism," and Anthony (1998:ix) calls it "the most striking thing to emerge in the U.S. environmental movement." Whether it developed in the environmental movement or from the civil rights movement (or both) is perhaps a moot point. Nevertheless, Cole and Foster call the civil rights movement "perhaps the most significant source feeding into today's environmental justice movement" (2001:20).

It will be useful at this point to examine the allied and earlier concept of environmental equity. An observer of even the least critical of news reports cannot fail to realize that environmental risks and hazards do not affect everyone to the same degree, both in the United States and abroad. The geographical distribution of risk and inequity is uneven within any state (Morello-Frosch 1997; Faber and Krieg 2002; Commonwealth of Massachusetts 2002), within the United States (Boyce et al. 1999), and between one country and another (Torras and Boyce 1998). Cutter (1995:113) notes that "environmental equity is a broad term that is used to describe the disproportionate effects of environmental degradation on people and places." Heiman (1996:114) adds that "environmental justice demands more than mere exposure equity. . . . it must incorporate democratic participation in the production decision itself." Environmental justice is a more targeted concept than environmental equity. As Bullard's (1994) "environmental justice framework" shows, it has at its heart the notion of righting a wrong, correcting an unjustly imposed burden, whereas environmental equity typically focuses on sharing burdens equally.

Environmental justice was, in its earliest and most specific sense, aimed at people of color (Bullard 1994; Epstein 1997), although, like Pulido (1996a), Epstein (1997:80) notes the "cross- or inter-racial nature of environmental injustices." This may be the case in terms of injustices, but it has not eased activist racial tensions in specific EJ cases such as those in Louisiana, detailed by Timmons Roberts and Toffolon-Weiss (2001), nor in the wider movement as a whole, tensions that "hearken back to the 1990 letter written by environmental justice activists to the 'Big Ten' national environmental organizations accusing them of not truly representing the interests of communities of color" (ibid.:57). In another vein, Pulido (1994:17) notes that "a distinct but prominent sub-movement is being formed that is limited to people of color." Based on my observations at the Second National People of Color Environmental Leadership Summit in Washington, D.C., in October 2002, I would say that within the "people of color" environmental justice movement, there are several race-based factions that hinder progress on many of the pressing issues that the movement as a whole should address and that are the subject of this book.

Pulido (1996b), like Cutter (1995), has challenged the monolithic view of environmental justice and the environmental justice movement. She bravely wades into the debate on the meaning of *racism* inherent in environmental racism research, arguing for a more differentiated concept of race: "the racism experienced by . . . Asian women in corporate America and that experienced by undocumented Mexican immigrants" is different and "the discourse of racism . . . can simultaneously serve to silence other visions, interpretations and experiences" (152). In effect, she is arguing against environmental justice as a unitary racial project that can "block the consideration of equally liberating racial projects and discourses" (154). This valuing of difference is similar to Schlosberg's (1999) call for an understanding of the diversity of ideas and difference in the movement and especially to his point that "rather than one particular frame or ideology, there is a coexistence of multiple political beliefs as to the causes, situation of, and possible solutions for issues of environmental justice" (111).[4]

So, because of its increasingly broad usage, especially outside the United States (Costi 1998; Agyeman 2000; Adeola 2000; Friends of the Earth Scotland 2000; Dunion and Scandrett 2003), *environmental justice* will be used in this book to include poor and disadvantaged white groups as well as people of color. As Cutter (1995:113) notes, "environmental justice . . . moves beyond racism to include others (regardless of race or ethnicity) who are deprived of their environmental rights, such as women, children and the poor."

In October 1991, the First People of Color Environmental Leadership Summit was held in Washington, D.C. Attracting more than six hundred delegates from fifty states, the main outcome of the summit was the Principles of Environmental Justice. These are a set of seventeen criteria around which to develop and evaluate policies for environmental and social justice. In October 2002, at the Second National People of Color Environmental Leadership Summit, the focus was on ratifying the Principles of Environmental Justice, planning the future direction of the movement, developing a series of policy papers on major environmental justice topics, and, most significant from the perspective of this book, looking at ways of working with the wider environmental movement (Summit II Executive Committee 2002).

At the federal level, there is an Office of Environmental Justice in the EPA and a National Environmental Justice Advisory Council (NEJAC). In 1994, former president Clinton's Executive Order 12898 on environmental justice reinforced 1964 Civil Rights Act Title VI, which prohibits discriminatory practices in programs receiving federal funds, and directed all federal agencies to develop policies to reduce environmental inequity. However, a study by a panel of the National Academy of Public Administration (NAPA 2001a) concluded that the EPA needs to be more proactive in integrating environmental justice into its core mission. Although the EPA has been trying for ten years to ensure that its permitting programs achieve fair treatment and meaningful involvement of all people, the study found that it has yet to effectively incorporate environmental justice issues into its permits.

More recently, the United States Commission on Civil Rights (2003) reported that progress toward fully implementing Executive Order 12898 has been patchy, with EPA, HUD, DOT, and DOI failing to do so. More damning still, the Office of the Inspector General's Evaluation Report (2004:i) notes bluntly that "EPA has not fully implemented Executive Order 12898 nor consistently integrated environmental justice into its day to day operations." Critically, however, the report goes on to say that "EPA has not identified minority and low-income populations addressed in the Executive Order, and has neither defined nor developed criteria for determining disproportionately impacted." This lack of identification criteria and standardization has caused problems for regional offices that want to develop policies and for states like Massachusetts that are looking to the EPA for guidance on policy development. In the end, many states had to go it alone.

However poorly implemented, Executive Order 12898 symbolically heralded the historic shift of the movement's claims and discourse from the Bible Belt to the Beltway, or, as Goldman (1996:131) puts it, "from street-level protests to federal commissions." Cole and Foster (2001: 161) call this "a profound institutional transformation on the widest scale." However, and crucially, as Schlosberg (1999:108) notes, "rather than create large, Washington based, bureaucratic organizations exemplified by the Big Ten,[5] concerned and active citizens have created a number of grassroots environmental networks" such as the Indigenous Environmental Network, the Northeast Environmental Justice Network, and the Southwest Network for Economic and Environmental Justice. Cole and Foster (2001:132) argue that "the establishment of broad-based social justice networks dispels the notion that the Environmental Justice Movement is simply another example of 'NIMBYism.' . . . it is engaged in something much more transformative."

In part as a result of the success of these networks in supporting other struggles, the 1990s saw real gains for the EJ movement including those in Chester, Pennsylvania (against Pennsylvania Department of Environmental Protection); South Central Los Angeles (against the LANCER mass-waste incinerator); Claibourne, Louisiana (against the Louisiana Energy Services uranium enrichment facility); St. James, Louisiana (against Shintec Inc. and their proposed chemical plant); and Kettleman City, California (against Chemical Waste Management).

While EJ issues, organizations, and networks do not necessarily follow political and administrative boundaries, some states have responded positively to the EJ agenda, often because of the election of sympathizers or the culturing of sympathizers already in office. An example of the former would be Massachusetts state senator Jarrett T. Barrios, many of whose supporters and campaign workers were EJ activists in the metro Boston towns of Chelsea and Everett. An example of the latter is Massachusetts state senator Diane Wilkerson, a longtime supporter of ACE, who has become a mainstay of the Massachusetts EJ community. Wilkerson, with the support of ACE, proposed the Environmental Justice Designation Bill (S.1060), a powerful new tool that (if successful) would have directed state officials to designate "Environmental Justice Populations" (discussed later in this chapter) as Areas of Critical Environmental Justice Concern (ACEJC), modeled on the state's current law on Areas of Critical Environmental Concern, which protects natural resources.

In summary, as Agyeman and Evans (2004:155–156) argue, "environmental justice may be viewed as having two distinct but inter-related dimensions. It is, predominantly at the local and activist level, a vocabulary for *political opportunity, mobilization and action*. At the same time, at the government level, it is a *policy principle* that no public action will disproportionately disadvantage any particular social group."

Some Policy Tools

In addition to Section 101 of the National Environmental Policy Act (NEPA), Title VI of the Civil Rights Act, Executive Order 12898, and the general discretion of an administrative agency, there are many policy tools that can be used by advocates and activists.[6] These tools include the Clean Air Act; the Clean Water Act; the Resource Recovery and Conservation Act; Comprehensive Environmental Response; the Recovery and Liability Act (Superfund); the Federal Insecticide, Fungicide, and Rodenticide Act; the Federal Food, Drug, and Cosmetic Act; the Safe Drinking Water Act; and the Toxic Substances Control Act.

There is also the 1986 Emergency Planning and Community Right-to-Know Act (EPCRA), which was designed to help local communities protect public health, safety, and the environment from chemical hazards. The Toxics Release Inventory (TRI) comes under Section 313 of EPCRA. Users of chemicals including electrical utilities, metallic and coal mining operators, commercial hazardous-waste treatment facilities, chemical and allied products manufacturers, solvent recovery services, and petroleum bulk terminals are required to report to the EPA the levels of listed chemicals that are released to air, water, or land. At present, there are more than six hundred chemicals on the list, from a start of three hundred in 1987.

A more deliberative tool that is being used by the environmental justice movement is the Good Neighbor Agreement, a voluntary, legally binding agreement between an industry and the community, which is used in some states to improve accountability. The agreement includes clauses on community access to information, negotiated improvements in pollution prevention, local job guarantees, and other local economic benefits. Similarly, a Community Benefits Agreement is a legally binding contract between a developer and a community in which benefits are provided by the developer in return for community support for a project. A newer tool with abundant promise to level the pro-industry playing field in pluralistic decision-making processes is the Precautionary Principle,[7] which is based on the German *Vorsorgeprinzip,* or Foresight Principle. The Precautionary Principle states that when an activity raises the threat of harm to human health or to the environment, precautionary measures should be taken even if cause and effect relationships are not fully established scientifically.

The biggest U.S. success of the Precautionary Principle to date has been at the city level—the eight-to-two vote by the City of San Francisco Board of Supervisors to adopt it as a citywide environmental ordinance in June 2003 was the first in the nation. San Francisco's proactive policy states, "the City sees the Precautionary Principle approach as its policy framework to develop laws for a healthier and more just San Francisco" (City of San Francisco 2003). The boost for communities, especially low-income communities and those of color that bear a disproportionate burden of environmental, health, and other risks, is that the Precautionary Principle has the ability to shift the legal burden of proof to the proponent of the risk. In other words, instead of communities undertaking the unwieldy task of proving after the fact that harm is being done to them, industry must prove in advance of the risk that their proposed activity is harmless. This approach could, if fully implemented and enforced, support community residents who are often outspent and out-finessed by the legal machines and deep pockets of big industry. With San Francisco's new ordinance in place, environmental justice activists, residents, and public health experts have witnessed the tables turning. While traditional risk assessment approaches ask "how *much* environmental harm will be allowed?" in San Francisco policymakers will ask a very different question, "how *little* harm is possible?"

The Precautionary Principle may well begin to shape policy more widely in the United States, as it does in the European Union, if the Be Safe coalition has its way. Advocating a "better safe than sorry" approach, the coalition includes the Center for Health, Environment, and Justice (CHEJ), *Rachel's Environment and Health News,* the Environmental Health Alliance, and a diverse group of national, state, and grassroots organizations. One other area of expanding policy interest for the environmental justice movement is the issue of clean production, which is more fully dealt with in chapter 4.

Democratizing Risk and Research

A major policy-level achievement of the EJ movement has been its critique of expert-led processes in both risk assessment and research and its ability, with those in allied movements such as health, to shape more transparent, accountable, and democratically informed processes.

Brulle (2000:208–209) argues that

> the discourse of environmental justice sees the use of scientific experts as part of a system of oppression and domination. Without access to experts of their own, some local community activists see scientific discussions as a means of keeping their viewpoints and concerns from being addressed by government officials. As a consequence, environmental justice groups challenge the authority of scientific experts to adequately express community concerns.

Regarding risk assessment, as Goldman (2000:541) puts it, "environmental justice is a risk analyst's worst nightmare." If people of color and those with low incomes are suffering a disproportionate burden of environmental bads, and an expert who is part of the system of oppression and domination comes along, as Brulle (2000) argues, and wants to use what is perceived to be a tool from that unjust system, communities are not going to be very happy. The problem is, as Fischer (2002: 127–128) states, "risk assessment's dilemma is rooted in its technical orientation to risk. The difficulty is lodged in the treatment of risk as a relatively structured problem that can be approached in a relatively uncomplicated and straightforward quantitative fashion." Citizen perceptions of risk are far richer and more qualitative and varied than expert perceptions. Using Plough and Krimsky's (1987) concepts of "technical rationality" (faith in science and empiricism) and "cultural rationality" (faith in personal and community experience) and drawing on the 1996 National Academy of Sciences report *Understanding Risk: Informing Decisions in a Democratic Society* (Stern and Fineberg 1996), Fischer (2002:247) calls for a culturally informed participatory alternative to expert-led risk assessment that uses "an 'analytic-deliberative method' capable of bringing together citizens and experts." Fischer's argument is that the *analysis* part of the method is the expert part based on rigor and protocol but that the *deliberation* part "informs risk decisions, such as deciding which harms to analyze and how to describe scientific uncertainty and disagreement. Appropriately structured, deliberation contributes to sound analysis by adding knowledge and perspectives that improve understanding, and contributes to the acceptability of risk characterization by addressing potentially sensitive procedural concerns" (247–248). Goldman (2000) calls this "Community-Led Risk

Analysis." He continues, "it is slow, interactive, educational, empowering. It uses some techniques developed by the experts and subverts others. Its products are often unexpected: including concerns about mold in public housing as well as toxic dumps, or auto-body shops in addition to lead poisoning, or fish consumption; and it violates many rules of positivistic science, especially involving the separation of researchers from their subjects" (544).

Regarding research, in 1994, the National Institute of Health Sciences facilitated an interagency conference called "Health Research Needs to Ensure Environmental Justice." According to Shepard et al. (2002), the conference, "attended by over 1000 persons, 400 of whom were environmental justice advocates . . . resulted in an expressed appreciation of the importance of community involvement in setting and implementing research agendas to address environmental justice issues" (139). Community-based participatory research (CBPR) fulfills Principles of Environmental Justice item 16 to the letter in that it is "based on *our* experience and an appreciation of *our* diverse cultural perspectives" (my emphasis).

What is CBPR? According to the Johns Hopkins Urban Health Institute,[8] it is a *process* that involves community members or recipients of interventions in all phases of the research process, including (a) identifying the health[9] issues of concern to the community; (b) developing assessment tools; (c) collecting, analyzing, and interpreting data; (d) determining how data can be used to inform actions to improve community health; (e) creating the research designs; (f) designing, implementing, and evaluating interventions; and (g) disseminating findings.

Shepard et al. (2002:139) build on this definition. First, "scientists[10] work in close collaboration with community partners involved in all phases of the research, from the inception of research questions and study design, to the collection of the data, monitoring of ethical concerns, and interpretation of the study results." Second, "in CBPR, the research findings are communicated to the broader community—including residents, the media and policymakers—so they may be utilized to effect needed changes in environmental and health policy to improve existing conditions." Third, CBPR "seeks to build capacity and resources in communities and ensure that government agencies and academic institutions are better able to understand and incorporate community concerns into their research agendas."

The Principles of Environmental Justice

How has this grassroots movement, full of community outrage, gained such strength and influence in a relatively short time? The federal and state apparatus, based on the policy principle that no public action should disproportionately disadvantage any particular social group, have clearly played a part, stronger in some states than in others, as have the adoption and use of different tools by activists, the use of different repertoires of action, the democratization of risk and research, and careful discursive framing(s) and resource mobilization techniques (see, for instance, Novotny 2000; Taylor 2000). Two pivotal acts were the problematization of the word *environment* and the development of the Principles of Environmental Justice.

The grassroots redefinition of environmental issues to include not only wildlife, recreational, and resource issues but also issues of justice, equity, and rights gave birth to the environmental justice movement. In so doing, *environment* became discursively different: it became an issue not just for the Sierra Club, National Wildlife Federation, and National Audubon Society but also for the civil rights movement. This linkage between a redefined *environment* ("where we live, where we work and where we play") and a social-justice analysis from the civil rights movement produced the dynamic movement in evidence today.

The common perception of environmental justice and its activists, and the movement itself, as being solely concerned with stopping toxic incursions into low-income neighborhoods and neighborhoods of color is firmly quashed on reading the Principles of Environmental Justice. Although the statement of principles was developed in 1991 and has served as the mission, mantra, and mandate for the movement, there has not, according to Taylor (2000:537), "been a systematic analysis of its content to examine the ecological and social justice components of the statement." This in part reflects a major cleavage within the environmental justice movement between activists and academics.[11]

Many movement activists, often those in small neighborhood or community groups, are largely reactive in the face of real day-to-day threats to their communities. As I made clear in the introduction, this is not what the Principles of Environmental Justice advocate, but reaction is reality for many communities starved of resources because, according to Cole and Foster (2001), they are "largely, though not entirely, poor or working-class people. Many are people of color who come from communities that are disenfranchised from major societal institutions" (33) and who "enter the decision-making process with fewer resources than other interests in the decision-making process" (109). They are the visible face of the direct-action wing of the environmental justice movement: the marchers, the activists, and the protesters who get attention in the local and sometimes national media. This is not, however, a criticism of the local activists. Given their resource levels, they achieve a tremendous amount in neighborhoods and communities under threat. Their focus has to be stopping threats now, in the hope of one day being able to take the time to envision a future that they will take an active role in shaping.

Although a generalization, the movement's larger not-for-profits and university centers, such as the Environmental Justice Resource Center at Clark Atlanta University or the Deep South Center for Environmental Justice at Xavier University in New Orleans, primarily want the movement to be more proactive, working on systemic issues as ACE does (see chapter 5). These groups see reaction to external threats as a necessary but not sufficient strategy to take the movement to its next level. They see the need to respond to Foreman's (1998) critique that environmental justice should show what it is *for,* in policy and planning terms, not just what it is *against.* They see the Principles of Environmental Justice as "a well developed environmental ideological framework that explicitly links ecological concerns with labor and social justice concerns" (Taylor 2000:538).

I will argue and show through my Just Sustainability Index, fifteen vignettes, and case study of ACE that it is precisely this linkage between ecological/environmental and social justice concerns that creates the possibility of a relationship between environmental justice and sustainability. The nexus, however, is between environmental justice and the emergent JSP, not between environmental justice and nature/stewardship-oriented environmental sustainability, where there are far fewer possibilities for linkage or connection. The JSP represents both the ideological and the activist cutting-edge of the sustainability movement.

Definition

As I made clear in the introduction to this book, environmental justice, like sustainability, is a contested and problematized concept. Therefore, defining it is not an easy task. Like sustainability, there are many possible definitions. The Commonwealth of Massachusetts uses the following definition in its "Environmental Justice Policy":

> Environmental justice is based on the principle that all people have a right to be protected from environmental pollution and to live in and enjoy a clean and healthful environment. Environmental Justice is the equal protection and meaningful involvement of all people with respect to the development, implementation and enforcement of environmental laws, regulations and policies and the equitable distribution of environmental benefits. (Commonwealth of Massachusetts 2002:2)

This definition will inform the arguments made throughout this book. It has *procedural* justice aspects ("meaningful involvement of all people"), *substantive* justice aspects ("right to live in and enjoy a clean and healthful environment"), and *distributive* justice aspects ("equitable distribution of environmental benefits"). In line with arguments made in the previous section and throughout this book, it also makes the case that environmental justice should not only be *reactive* to environmental bads, important though this is, but that it should also be *proactive* in the distribution and achievement of environmental goods—for instance, in relation to this book, creating a sustainable community with a higher quality of life.

Framing Environmental Justice

Following the seminal work of Capek (1993) and Sandweiss (1998), Taylor (2000) has carefully analyzed the spectacular growth of the environmental justice movement over the past twenty years. She, like Sandweiss (1998), uses a social constructionist analysis (Spector and Kitsuse 1973; Hannigan 1995) that sees reality as socially rather than objectively constructed and that uses, among others, the concept of *framing*.

Sandweiss (1998:32) notes that framing "calls attention to a problem but also seeks to identify both its causes and possible remedies." More specifically, Entman (1993:52) argues,

> framing essentially involves selection and salience. To frame is to select some aspects of a perceived reality and make them more salient in a

> communicating text, in such a way as to promote a particular problem definition, causal interpretation, moral evaluation, and/or treatment recommendation for the item described.

Novotny (2000) links framing to another facet of social movement mechanics: mobilization. He notes that

> framing is, in a very real sense, part of the repertoire of mobilization strategies that are available to a movement, so that the movement filters the problems it is confronting through the history, the beliefs, the language and the cultural experiences that are seen by its leaders as most likely to engender widespread sympathy and involvement (7).

This echoes Benford's (1993:693) point that "a movement's claims not only need to resonate with the experiences of its audiences, they must also correspond with the cultural narrations, stories, myths and folk tales of their culture." This "dramaturgical" approach is characterized by Benford and Hunt (1992:37) as going beyond rhetoric "to consider a plethora of additional processes associated with the social construction and communication of meaning, including formulating roles and characterizations, managing performance regions, controlling information, sustaining dramatic tensions and orchestrating emotions."

In other words, framing is a dramatic and politico-cultural device that develops a deep resonance among its intended audience and acts as a rallying cry, a call to action. However, framing and mobilization are but two factors involved in the emergence and development of social movements such as the EJ movement. A third factor is political opportunity, in which "the broad political environment in which the movement is embedded will . . . constitute a powerful set of constraints/opportunities" (McAdam et al. 1996:2).

Taylor (2000) argues that "the environmental justice 'frame' is a master frame that uses discourse about injustice as an effective mobilizing tool" (508). She continues that "master frames are styles of punctuation, attribution, and articulation . . . [that] can be viewed as crucial ideological frameworks akin to paradigms" (514). She develops this idea by arguing that what Best (1987) calls "claims makers" (NGOs, policy entrepreneurs, and professionals) make "claims" (complaints about social problems) that "use rhetorical motifs . . . recurrent figures of speech, such as *vanishing wildlife, toxic soup, Toxic Doughnut,* or

Cancer Alley, that amplify the problem and give added moral significance to the claim" (514).[12]

Deriving its inspiration from and linking itself to the civil rights movement, the environmental justice movement "appropriated . . . the preexisting salient frames of racism and civil rights" (Taylor 2000:62). This, Taylor argues, has led to the development of the EJP, which "is most clearly articulated through the Principles" (537) and is "the first paradigm to link environment and race, class, gender, and social justice concerns in an explicit framework" (542). When comparing the EJP to the NEP, described as "a new set of [environmental] beliefs and values" (Milbrath 1989:118), Taylor notes,

> the EJP has its roots in the NEP, but it extends the NEP in radical ways. . . . The EJP builds on the core principles of the NEP; however, there are significant differences . . . vis à vis the relationship between environment and social inequality. The NEP does not recognize such a relationship; consequently it has a social justice component that is very weak or non-existent. (2000:542)

This is a key point that will be elaborated further in chapter 2 through a discussion of the most current expressions of the NEP,[13] namely, sustainability, sustainable development, and the development of sustainable communities.

Developing State Level Environmental Justice Policy

Following Connecticut's lead in 1993, some individual states are actively seeking to develop environmental justice policies[14] and laws. Massachusetts, for example, set up the Massachusetts Environmental Justice Advisory Committee (MEJAC) in 2000 and launched its policy in October 2002. I was on this committee, along with around ten others drawn from a range of occupations, including Penn Loh, executive director of ACE.

Modeled on the EPA's National Environmental Justice Advisory Council (NEJAC), the MEJAC organized all public outreach, held public meetings, visited neighborhoods, and attended presentations by activists from groups across the state in order to show policymakers a range of environmental justice interests and issues in communities at risk in Massachusetts. The MEJAC guided the development of the environmental justice policy and made recommendations for its implementation in the December 2000 Public Consultation Draft. In the second phase of policymaking, the State Working Group, made up of representatives employed by the Commonwealth, was charged with developing an implementation strategy based on that Public Consultation Draft.

A major issue MEJAC had to deal with early on was, if the Commonwealth was going to try to protect and target resources to those disproportionately affected by environmental bads, how to delineate those affected. In other words, the committee's first question was, What is an environmental justice community? Clearly, the EPA, with a string of critical reports (NAPA 2001a; U.S. Commission on Civil Rights 2003) but especially that of the Office of the Inspector General (2004), was not able to provide leadership in this critical area.

Eady (2003:172) asks two key questions:

> First, what percentage of a community must be "of color" in order to comprise a community of color or minority community? The Executive Order on Environmental Justice, 12898, signed by former President Clinton suggests that a minority population should be found "where . . . the minority population of the affected area exceeds 50 per cent."[15] Second, what borders define a community? Under the paradigm guiding many environmental protection programs, municipal boundaries define "community." In that case, many communities in the nation, including the City of Boston, do not meet the Executive Order 12898 threshold and [Boston] neighborhoods like Roxbury and North Dorchester would not be reached by environmental justice protections.

In earlier iterations of the Massachusetts Environmental Justice Policy, what are now called Environmental Justice Populations were called Environmental Justice Communities. This may reflect the perception among the Commonwealth's Executive Office of Environmental Affairs (EOEA), an agency of the newer Office for Commonwealth Development (OCD) formed in 2003, that to label specific communities Environmental Justice Communities could be detrimental or introduce stigma in terms of realty, insurance, or other fiscal issues. I must also make it clear that neither Environmental Justice Communities nor Environmental Justice Populations is synonymous with the Environmental Justice Movement. The former are *communities of place*: they are administrative designations that progressive states will need to develop if they are serious about moving from written policy to implementation, whereas the latter is a social movement, a *community of interest.*

After a protracted process, EOEA arrived at the following definition of Environmental Justice Populations:

> EJ Populations are those segments of the population that EOEA has determined to be most at risk of being unaware of or unable to participate in environmental decision-making or to gain access to state environmental resources. They are defined as neighborhoods (U.S. Census Bureau census block groups) that meet one or more of the following criteria:
>
> - The median annual household income is at or below 65 percent of the statewide median income for Massachusetts; or
> - 25 percent of the residents are minority; or
> - 25 percent of the residents are foreign born, or
> - 25 percent of the residents are lacking English language proficiency. (Commonwealth of Massachusetts 2002:5)

While imperfect, these criteria are a base around which to implement and evaluate the EJ policy. MassGIS, Massachusetts's Geographic Information System service, has now mapped all EJ Populations in the Commonwealth, based on currently available 2000 U.S. Census data.[16] The policy acknowledges that EJ Populations make up 5 percent of the Commonwealth's land area and include about 29 percent of its population. Regarding location, unsurprisingly,

> many of these Environmental Justice Populations are located in densely populated urban neighborhoods, in and around the state's oldest industrial sites, while some are located in suburban and rural communities. (Commonwealth of Massachusetts 2002:5)

Three weeks after Massachusetts released its environmental justice policy in a public comment draft (December 2000), Faber and Krieg (2002)[17] unveiled their long-anticipated report, "Unequal Exposure to Ecological Hazard: Environmental Injustices in the Commonwealth of Massachusetts." In a revealing look at EJ Populations in Massachusetts, they undertook an in-depth analysis of the socio-geographic distribution of hazards across 368 communities in the Commonwealth. They

TABLE 1.1
Most Intensively Overburdened Communities in Massachusetts in Total Points per Square Mile

Rank	Town name	Points per square mile	Class status of town	Racial status of town
1	Downtown Boston[a]	224.8	Low income ($29,468)	High minority (31.9%)
2	Charlestown	134.3	Medium-low ($35,706)	Moderate-low (5.1%)
3	Chelsea	127.4	Low income ($24,144)	High minority (30.3%)
4	South Boston	126.2	Low income ($25,539)	Low minority (4.2%)
5	East Boston	123.3	Low income ($22,925)	Moderate-high (23.6%)
6	Cambridge	115.0	Medium-low ($33,140)	Moderate-high (24.9%)
7	Somerville	104.7	Medium-low ($32,455)	Moderate-low (11.3%)
8	Roxbury	101.3	Low income ($20,518)	High minority (94.0%)
9	Allston/Brighton	100.0	Low income ($25,262)	High minority (26.9%)
10	Watertown	98.6	Medium-high ($43,490)	Low minority (3.8%)
11	Everett	98.1	Medium-low ($30,786)	Moderate-low (6.0%)
12	Boston (all neighborhoods)	84.0	Low income ($29,180)	High minority (37.0%)
13	Dorchester	81.3	Low income ($29,468)	High minority (50.7%)
14	Lawrence	59.3	Low income ($22,183)	Moderate-low (34.9%)
15	Malden	57.8	Medium-low ($34,244)	Moderate-low (10.1%)
Totals	15 towns		14 of the 15 most intensively overburdened towns are of lower income status status (less than $40,000)	9 of the 15 most intensively overburdened towns are of higher minority status (15% or more people of color)

[a] Downtown Boston encompasses Central Boston and Chinatown, Back Bay and Beacon Hill, the South End, and the Fenway/Kenmore neighborhoods.
Source: Faber and Krieg (2002:278).

looked at both income-based and race-based biases in the distribution of seventeen types of hazardous sites and industrial facilities. Using enforcement data and other public data from Massachusetts's own regulatory agencies, they developed a point system for measuring and ranking cumulative exposure to these sites and facilities, which showed that "hazardous sites and facilities are disproportionately located and concentrated in communities of color and working-class communities" (277) (see table 1.1).

According to Eady,[18]

> while the report surprised few in the environmental justice and regulatory worlds, it sent shockwaves through neighborhoods across Massachusetts where residents sensed an unfair pollution burden but did not know what to call it. All of a sudden, "environmental justice" was a

> widespread battle cry, not just among the state's communities of color and low-income neighborhoods, but all across the landscape. (2003:171)

The work of environmental justice groups in Massachusetts, including ACE and the networks that it services—such as the Greater Boston Environmental Justice Network (GBEJN), which brings together thirty neighborhood groups, and the Northeast Environmental Justice Network (NEJN)—was vindicated by the results of Faber and Krieg (2002). Indeed, their work, along with the publication of the Commonwealth's policy, provided a special moment in Massachusetts when EJ issues were at the fore, resulting in the two bills described below.

Into the Mystic

Nowhere was the sense of shock greater than in the communities within the Mystic River Watershed in the metro Boston area, where fully eight out of the fifteen most overburdened towns and cities in Massachusetts are to be found, according to the Faber and Krieg report (see table 1.1 and fig. 1.1: Charlestown, Chelsea, East Boston, Everett, Malden, and portions of Cambridge, Somerville, and Watertown).[19] Although Faber and Krieg's methodology and aims were different from those of the MEJAC, the Commonwealth also found these communities to have more than their "fair share" of EJ Populations.

The Mystic River Watershed encompasses an area of approximately seventy-six square miles, with twenty-one municipalities (in whole or in part) north and west of Boston (see fig. 1.1). The headwaters of the system begin in Reading and form the Aberjona River, which flows into the Upper Mystic Lake in Winchester. The Mystic River flows from the Lower Mystic Lake through Arlington, Medford, Somerville, Everett, Charlestown, Chelsea, and East Boston before emptying into the Boston Harbor. Main tributaries to the Mystic River include Mill Brook, Alewife Brook, Malden River, and Chelsea Creek (also known as the Chelsea River). The watershed contains forty-four lakes and ponds, the largest of which is Spot Pond in the Middlesex Fells, with an area of 307 acres.

Home to about 8 percent of Massachusetts's population (nearly half a million people) in less than 1 percent of its land area, the Mystic River Watershed is one of the most densely populated and urban watersheds in the state. The watershed includes relatively wealthy economically

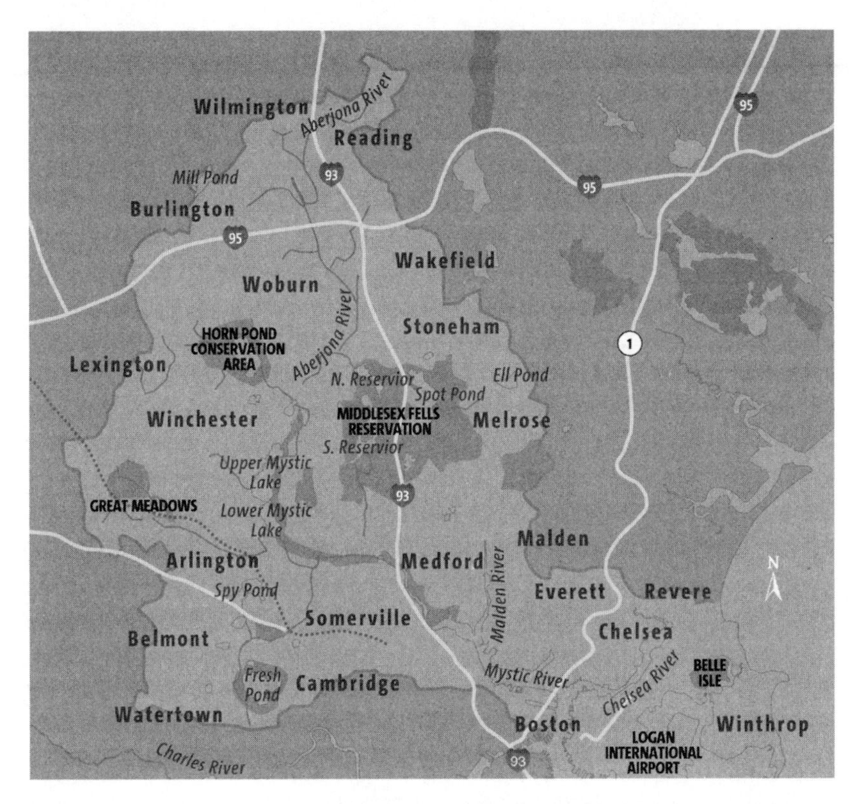

Fig. 1.1. The Mystic River Watershed (from the Mystic River Watershed Association).

developed suburbs such as Winchester and Stoneham as well as low-income urbanized-center communities such as Somerville and Chelsea (see fig. 1.1). The 1999 per capita income in these communities varied from $14,628 in Chelsea and $19,845 in Everett to $50,514 in Winchester and $46,119 in Lexington. According to the Massachusetts Department of Revenue, the range of median family income was between $32,130 in Chelsea and $49,876 in Everett to $111,899 in Lexington and $110,226 in Winchester.[20] As of September 2003, unemployment rates across Mystic communities indicate similar disparities. Not surprisingly, Chelsea and Everett residents endure a steeper rate of joblessness, 8.8 percent and 7.1 percent, respectively, while Lexington and Winchester citizens fare relatively better in securing a livelihood: 3.7 percent and 4.2 percent, respectively.

True to sprawling metropolitan development patterns most everywhere, population density varies across the watershed, thinning out with distance from the metro Boston core. In its urbanized-center communities, for example, approximately 35,080 Chelsea residents inhabit 2.19 square miles (land area), or 16,038 people per square mile, and in Somerville roughly 50,454 people reside in 2.67 square miles of land in the watershed (which is about 65 percent of the city's land mass), or 18,897 people per square mile. For the relatively well-off suburb of Stoneham, approximately 21,159 residents occupy 5.86 square miles, or 3,611 people per square mile, whereas in Lexington 6,100 people live in 3.21 square miles, or 1,900 people per square mile. Significantly, communities in the lower watershed (e.g., Chelsea, East Boston, and Somerville) are more ethnically diverse and have considerable numbers of recent immigrants. For instance, in the 2000 census, East Boston had 39 percent Hispanic, 3 percent black, 4 percent Asian or Pacific Islander, less than 1 percent Native American, 1 percent "other," and 3 percent multiracial, fully 50 percent of the population of 38,413.[21]

Development in the watershed began in the 1600s and has included extensive industrial and manufacturing facilities across the area's seventy-six square miles. This has resulted in the release of hazardous chemicals to soils, groundwater, and surface waters. At present there are two Superfund sites and several hundred state-identified hazardous-waste disposal sites in the watershed. The lower half of the watershed contains combined sewer overflows (CSOs) that degrade overall water quality by discharging untreated sewage into the Mystic River and the Alewife Brook during storms (Perez et al. 2002). Illicit connections of sanitary-waste pipes into storm-water drains contribute to inputs of sewage during dry weather. Excessive application of nutrients in urban and industrial landscaping practices, perhaps chief among the non-point-source pollution factors (Kirshen et al., 2000), affect water quality from runoff of storm water across impervious surfaces. In the Mystic watershed, only 17 percent is designated as open space, and in Somerville, 85 percent of the land is impermeable, making urban runoff a substantial problem. The large amount of organic matter in the river leads to generally low dissolved-oxygen levels, high turbidity, and high quantities of pathogenic bacteria. Industry in the lower watershed threatens both water and air quality standards. Although the towns of Everett and Chelsea are both mostly bounded by the Mystic and its tributaries —the Malden and Chelsea rivers, respectively—neither has appreciable

public access to the waterfronts. Since 1978, Chelsea's waterfront has been rezoned a designated port area (DPA) as part of the Boston Harbor, whereby development for industrial use is prioritized (Hynes 2003). Everett's waterfront to the Mystic is also a DPA (though not its waterfront to the Malden River); the Mystic along both East Boston and Charlestown was part of the 1978 DPA designation. As a result of industry and regional energy-distribution facilities and Logan International Airport cargo terminals, heavy commercial and industrial truck traffic impact residential neighborhoods throughout the lower watershed communities.

Implementing State-Level Environmental Justice Policy

What does the Commonwealth intend to do about the Mystic watershed and about the wider injustices in Massachusetts identified by both its own staffers, MEJAC and Faber and Krieg? It is working on two different levels, policy and law.

First, at the policy level,

> it is the policy of the EOEA that environmental justice shall be an integral consideration to the extent applicable and allowable by law in the implementation of all EOEA programs, including but not limited to, the grant of financial resources, the promulgation, implementation and enforcement of laws, regulations and policies, and the provision of access to both active and passive open space. (Commonwealth of Massachusetts 2002:4)

In real terms, this means that the Commonwealth will increase public participation and outreach through the development of strategies, training, fact sheets, and Regional EJ Teams; minimize risk to EJ Populations through targeted compliance, enforcement, and technical assistance; encourage investment and economic growth, particularly around contaminated sites; infuse state resources by developing an inventory of underutilized commercial/industrial properties; incorporate an environmental justice criterion in the awarding of technical assistance, grants, and audits in 21E (hazardous waste and brownfield) sites in EJ Populations; and promote cleaner production and the creation, restoration, and maintenance of open spaces.

While it is still too early to evaluate the effectiveness of this policy in comparison to other state policies, its rigor, intent, and apparent political support are encouraging.[22] However, there are three significant problems. First, the policy is, as the quote above shows, limited to "all EOEA programs." EOEA is an agency of the much bigger OCD, whose role is to "care for the built and natural environments by promoting sustainable development through the integration of energy, environmental, housing, and transportation policies, programs, and regulations" (Office for Commonwealth Development 2004). Does the policy not apply to the whole OCD and its policies and programs? In other words, are environmental justice considerations not part of sustainable development in Massachusetts? Second, EOEA's current full-time environmental justice officials lack the clout to do anything other than advise and guide action regarding the initiatives described above. Third, EOEA has, in my opinion, missed a great opportunity by setting the bar too low. It aims to bring EJ Populations (5 percent of the Commonwealth's land area, 29 percent of its population) only up to the level of other (unsustainable) Massachusetts communities. One could argue that the array of strategies, investments, enforcements, and promotions proposed will move EJ Populations toward greater sustainability. However, this goal is neither implicit nor explicit in the policy. Indeed, some of the targeted economic growth could work against sustainability.

Massachusetts does not have a policy on sustainable communities as such. However, it has two initiatives which show some potential. One is Executive Order 438 (2002),[23] which took an approach to growth management by creating a grassroots, municipally driven smart-growth initiative called Community Preservation. Among other topics, Community Preservation encompasses land and watershed protection, affordable housing, historic preservation, economic development, and transportation. The other initiative is the development by the OCD, in 2004, of a set of ten sustainable development principles.[24] In the context of this chapter, it is pleasing to see that the OCD's third principle, "Be Fair," mentions environmental justice. However, it is not immediately obvious how EOEA's Environmental Justice Policy and Community Preservation Act and OCD's Sustainable Development Principles relate to each other in practical, implementation terms. A new OCD capital-spending program for FY 2005 called Commonwealth Capital prioritizes projects that reflect the Sustainable Development Principles and that show partnerships with municipalities, yet, other than the appearance of the word "brownfield," the program gives no specific criteria related to the Commonwealth's environmental justice policy.

Massachusetts, it seems, has the components and the policy language, but, as we shall see in the next chapter, this policy must also be "joined up," that is, related to other policy areas. A more promising and integrative approach, described more fully in chapter 3, is being taken by the State of Maryland through its Commission on Environmental Justice and Sustainable Communities. By linking these two issues through a commission, the connection between the two should be clearer than if there were two separate commissions.

Second, at the legal level, state senator Jarrett T. Barrios has sponsored an EJ bill (S. 190), *The Clean and Healthy Communities Act.* Senator Barrios is joined by Representative David Sullivan, who has introduced to the House *An Act to Promote Environmental Justice in the Commonwealth* (H. 2112). The bills would require

> the state to develop statewide regulations that give communities greater protections from even more pollution, and would establish a procedure under which additional communities that do not fall under direct demographic definition of an EJ Population may petition for such status. (From promotional literature)

Conclusion

I do not mean to suggest that the Massachusetts EJ model is unworkable in moving toward more sustainable communities, just that the state missed the opportunity to *specifically* and *explicitly* link the Commonwealth's two main sustainability policies—namely, the Community Preservation Act and the Sustainable Development Principles—to its own EJ policy despite policy integration being one of OCD's main roles.

A task for activists, groups, and academics around the United States should be to identify and map EJ Populations in their states, using at least as stringent criteria as those of Massachusetts. A National Academy of Public Administration report (2001b) argues that state agencies should develop performance, outcome, and accountability measures to reduce community exposures to environmental hazards. The panel studied legislation, policy, procedures, and tools that the states of Indiana, Florida, New Jersey, and California have used to address environmental justice issues, but it argues the approaches have produced few tangible improvements for communities of color and low income.

Such performance, outcome, and accountability measures, if targeted to defined areas similar to Massachusetts's EJ Populations, together with the adoption of the Precautionary Principle, could form the basis for arguments that remedial actions and improvements to right environmental injustices in such communities should go further than just "bringing them up to scratch" (i.e., up to our current state of *unsustainability*). If this does not happen, and if wealthier areas around them become more sustainable communities, EJ Populations may come to be seen as *people of color and poor people's sustainable communities.* As Roseland (1998:2) astutely argues, "sustainable communities are not merely about 'sustaining' the quality of our lives—they are about improving it." The path from environmental justice communities (or populations) to sustainable communities is indeed a long one.

2

The Sustainability Discourse and Sustainable Communities

What are sustainable communities? How do we define them, and what are their characteristics? If developing sustainable communities will improve the quality of our lives, how do we create them? In this chapter, I focus on four themes that help differentiate *environmental* sustainability, or the NEP, from *just* sustainability, or the JSP. First, I look at the origins and the theoretical and practical aspects of sustainability, sustainable development, and sustainable communities, focusing on a critique of environmental sustainability. Second, I look at a few policy tools and policies currently available and being used in U.S. cities and in cities and regions elsewhere in the world, focusing on San Francisco's attempts to integrate environmental justice into its sustainable development policy. Third, I look at the characteristics of a sustainable community. And fourth, I look at the dominant sub-national environmental policy discourse in the United States, civic environmentalism, as it relates to sustainable communities, differentiating between *narrow-focus* and *broad-focus* civic environmentalism. The former is firmly within the NEP; the latter, within the JSP and EJP.

Sustainable Development Theorizing

Around the same time as environmental justice was developing as a public policy issue, the ideas of *sustainability* and *sustainable development* were achieving prominence among local, national, and international policymakers and politicians, together with policy entrepreneurs in NGOs. Since the 1980s, there has been a massive increase in published and online material dealing with sustainability and sustainable development. This has led to competing and conflicting views of what the terms mean, what is to be sustained, by whom, for whom, and what is the most desirable means of achieving this goal.

To some, the sustainability discourse is too all encompassing to be of any use. To others, the words are usually prefaced by "environmental" and "environmentally," as in "environmental sustainability" or "environmentally sustainable development." *Environmental modernization,* the dominant policy-based discourse of sustainable development in Europe, is, according to Smith (2003:4),

> a discourse of eco-efficiency. Its primary concern is the efficient use of natural resources within a capitalist framework (Hajer 1995; Christoff 1996; Gouldson and Murphy 1997). Criticisms have been leveled at the lack of attention paid to social justice (both within and between nations) and the failure to conceive of nature beyond its value as a resource.

To still others, the discourse offers a sense of integrity and holism that is lacking in contemporary, reductionist, silo-based policymaking. Indeed, the European trend is to talk of sustainable development policymaking as "joined up" or "connected" policymaking, that is, policymaking in specific areas—for example, housing, economic development, environmental justice, or environment—with an eye to its effect on the policy architecture as a whole as opposed to the policy silo from which it came. This is not currently the case in Massachusetts, where, as we saw in the previous chapter, there are policies for environmental justice and sustainable communities/community preservation, and there are principles of sustainable development, originating from agencies in the same department (OCD), which even with a substantial capital program behind them do not appear to be fully joined up.

One thing I am increasingly sure of is that the science of sustainability is not our greatest challenge. In almost all areas of sustainability, we know scientifically what we need to do and how to do it; but we are just not doing it. An advertisement in the *New York Times,* paid for by outofgas.com, said the same: "It's time to free ourselves from foreign oil, and create millions of new jobs in the process. This is no pipe dream. The research and technology exist. We have the national wealth. Do we have the will?" (*New York Times* 2004:A9). Sustainability is not alone in this respect. There are other examples in society where the science of an issue runs way ahead of public and political discourse, including stem cell research and euthanasia.

This gap between *knowing* and *acting* is what the Real World Coalition, an alliance of leading UK social movement organizations (SMOs), calls "the sustainability gap" (Christie and Warburton 2001). As Brulle (2000:191) argues,

> with the exception of Commoner, the vast majority of ecological scientists have not examined the social and political causes of ecological degradation (B. Taylor 1992:133–151). While the natural scientists may have great competence in their specific areas of expertise, their social and political thinking is "marred by blindness and naivete" (Enzensberger 1979:389).

This is, unsurprisingly, not a view shared by the National Academy of Sciences (1999:2), which argues,

> the political impetus that carried the idea of sustainable development so far and so quickly in public forums has also increasingly distanced it from its scientific and technological base. As a result, even when the political will necessary for sustainable development has been present, the knowledge and know-how to make some headway often have not.

The problem is that the NAS is equating knowledge and know-how exclusively with *science and technology* knowledge and know-how. My experience in local government in the United Kingdom is that where politics is not the limiting factor, it is rarely the science or technology of sustainability that is. Indeed, the limiting factor is far more likely to be the *social science,* such as ensuring broad participation or communication with a diverse array of stakeholders. The sustainability gap presents us with the millennial sustainability challenge, which I believe is political and (public) attitudinal. Or, as Prugh et al. (2000:5) put it, "sustainability will be achieved, if at all, not by engineers, agronomists, economists and biotechnicians but by citizens."

Is sustainability only about "green issues"? There are differing views on this. Bossel (1998:xi), for instance, argues that "it is an interdisciplinary hodgepodge of environmental science, ecology, systems science, computer science, mathematics, and ecological economics." In the previous chapter, I argued that *sustainability,* with its action-oriented variant, *sustainable development,* is the most current expression of the NEP. This is because environmentalism and environmentalists are the premier drivers of much of the current sustainability project, hence its orientation as *environmental* sustainability. Similarly, I have argued elsewhere that

> sustainability . . . cannot be simply a "green," or "environmental" concern, important though "environmental" aspects of sustainability are. A truly sustainable society is one where wider questions of social needs and welfare, and economic opportunity are integrally related to environmental limits imposed by supporting ecosystems. (Agyeman et al. 2002:78)

This *just* perspective on sustainability is a view shared by most thinkers in the EJ movement and is more fully explored in later chapters. Mencer Donahue Edwards, former executive director of the Panos Institute in Washington, D.C., wrote an influential paper in the *EPA Journal,* called "Sustainability and People of Color," in 1992, around the time of UNCED. In it, he argued, as did my interviewees at ACE in chapter 5, that people of color *do* embrace sustainable development because it will lead to a United States "transformed by the guiding principles of freedom, justice and equality" (Edwards 1992:51). He also makes the point that the preamble to the Principles of Environmental Justice (see Appendix) agrees with his assessment:

> to begin to build a national and international movement of all peoples of color to fight the destruction and taking of our lands and communities, [we] do hereby reestablish our spiritual interdependence to the sacredness of our Mother Earth; to respect and celebrate each of our cultures, languages, and beliefs about the natural world and our roles in healing ourselves; to insure environmental justice; to promote economic alternatives which would contribute to the development of environmentally safe livelihoods; and to secure our political, economic, and cultural liberation . . .

Edwards believes that this passage is actually a manifesto for sustainable development, focusing as it does on social, economic, and environmental issues.

Building on this idea, I fully endorse four key points on the "greenness" of sustainability: First, Polese and Stren's (2000:15) simple argument that "to be environmentally sustainable, cities must also be socially sustainable." Second, that of Middleton and O'Keefe (2001:16): "unless analyses of development [local, national, or international] . . . begin not with the symptoms, environmental or economic instability, but with the cause, social injustice, then no development can be sustainable." Third, that of Hempel (1999:43): "the emerging sustainability ethic may be more interesting for what it implies about politics than for what it promises about ecology." Finally, that of Adger (2002:1716):

> I would argue that inequality in its economic, environmental, and geographical manifestations is among the most significant barriers to sustainable development. It is a barrier because of its interaction with individuals' lifestyles and because it prevents socially acceptable implementation of collective planning for sustainability.

In other words, sustainability is at least as much about politics, injustice, and inequality as it is about science or the environment.

Sustainability is interpreted in this book as "the need to ensure a better quality of life for all, now and into the future, in a just and equitable manner, while living within the limits of supporting ecosystems" (Agyeman et al. 2003:5). It represents an attempt to look holistically at the human condition, at human ecology, and to foster joined up or connected—rather than silo-based or piecemeal—policy solutions to humanity's greatest problems. The definition focuses on four main areas of concern that are the foundations of the JSP: quality of life, present and future generations, justice and equity in resource allocation, and living within ecological limits. Sustainable development is therefore a political and policy framework for improving the way we live, the way we distribute goods and bads and the way we do business on this planet of finite resources. At its heart, though, is a fundamental acknowledgment of social injustice as the root of our current unsustainability. The *sustainability transition,* from where we are now to where we need to go, should be accompanied by both an increase in equity and justice and an increase in environmental quality.

Before we lapse into a false sense of security regarding what growing numbers of researchers see as the pivotal nature of justice and equity in sustainability formulations, note that the single most frequently quoted definition of sustainable development does not explicitly mention social justice or intragenerational equity. This definition comes from the World Commission on Environment and Development (1987), also known as the Brundtland Commission, which argued that "sustainable development is development that meets the needs of the present without compromising the ability of future generations to meet their own needs" (43). This definition does, however, imply intergenerational equity and an important shift away from the traditional, conservation-based usage of the concept, as formed by the International Union for the Conservation of Nature's (IUCN) 1980 World Conservation Strategy, to a framework that emphasizes the social, economic, and political context of "development." By 1991, the IUCN had modified its definition: "to improve the quality of life while living within the carrying capacity of ecosystems" (IUCN 1991:9).

Unlike the definition of sustainability, quoted earlier, of Agyeman et al. (2003), neither the WCED or IUCN definitions specifically mentions justice or equity, which many commentators now hold to be of fundamental importance in developing sustainable communities and futures (see, for instance, Edwards 1992; Raskin et al. 1998; Haughton 1999; Athanasiou n.d.; Middleton and O'Keefe 2001; Hempel 1999; Campbell 1996; Low and Gleeson 1998; Ruhl 1999; Prugh et al. 2000; Adger 2002, Agyeman et al. 2003; McLaren 2003; Roberts 2003; Glasmeier and Farrigan 2003; Buhrs 2004). Of these, Buhrs (2004:434) is perhaps most direct: "addressing environmental justice issues is important, if not a precondition, for the achievement of global sustainability." The omission of *justice* or *equity* from the most frequently used definitions of sustainability adds weight to Taylor's (2000:542) point from the previous chapter that "the NEP does not recognize such a relationship [between environment and social inequality] . . . consequently it has a social justice component that is very weak or non-existent." I call this the *equity deficit* of environmental sustainability.

Agenda 21

The 1992 UNCED in Rio de Janeiro, more popularly known as the Earth Summit, boosted sustainability and sustainable development to their current status in policymaking circles among all levels of government around the world (see Campbell 1996). The major policy outcome of the Earth Summit was Agenda 21, a global agenda for sustainable development in the twenty-first century, which was adopted by more than 178 governments but, significantly, not by the U.S. government.

The Commission on Sustainable Development (CSD) was created in December 1992 to ensure effective follow-up of the Earth Summit and to monitor and report on the implementation of Agenda 21 at the local, national, regional, and international levels. The successor to the Earth Summit was the 2002 WSSD in Johannesburg.

Partially in reaction to the difficulty of accomplishing the task of sustainable development on a global scale, partially in response to the local manifestation of social, economic, and environmental problems, and partly because of the subsidiarity principle (local governments are the level of governance closest to the people), many local governments and their communities are now adopting more-transparent decision-making processes, sharing control and adopting the principles of sustainability at the local level. This institutional (mini) change meant that, by February 2002, there were 6,416 local governments in 113 countries that had either made a formal commitment to the principles of Local Agenda 21 (LA21)[1] or were actively undertaking the process (ICLEI 2002a; see also Southey 2001). Following the relative worldwide success of LA21 (especially when compared with national and supra-national efforts to move toward sustainability), together with other municipal and community partnership initiatives such as Cities for Climate Protection (ICLEI 2002b), a new, local movement for sustainable communities is striving to link the goals of economic vitality, ecological integrity, civic democracy, and social well-being, while ensuring a high quality of life for all (Hempel 1999; Roseland 1998).

A result of the flurry of sustainability-related rhetoric, if not activity, has been a steep learning curve for governments around the world at all levels. As Jacobs (1999:29) argues,

> the sustainable development discourse has set off a process of institutional learning. Throughout international agencies and national governments—and to a lesser extent, the business sector and other sectors—the sustainable development discourse is pushing institutions to reappraise their policies and policy making processes.

In this task, of course, they are assisted by a growing number of consultants and other experts. If, however, we understand that this current institutional learning pertains to the dominant discourse of sustainability, namely, environmental sustainability, then imagine the task of getting institutions to first understand just sustainability, which "represents an ideological challenge to current politics, not just a policy one" (Jacobs 1999:44), and then to practice it.

Institutional Learning in Reverse: The U.S. Government and Sustainability

Institutional learning in reverse may be the best way to describe the current U.S. government's approach to sustainability, which has gone from a promising approach during the days of the Clinton-Gore President's Council on Sustainable Development (PCSD) to a poor one, if the evidence in *Working for a Sustainable World: U.S. Government Initiatives to Promote Sustainable Development,* published for the WSSD by the Bush administration (USAID 2002), is correct.

Paula Dobriansky, U.S. under secretary, global affairs, announced at a packed press conference at the WSSD in August 2002 that "the United States is the world's leader in sustainable development. No nation has made a greater contribution and a more concrete contribution to sustainable development." Let me say at the outset that this statement is actually *true.* In terms of *funding,* the United States is (or will be) the world's leader in sustainable development, and this reflects the Bush administration's approach to sustainable development.

What Dobriansky's statement alluded to was the unveiling of five funding partnerships, or in WSSD jargon, Type II initiatives for sustainable development:

- The Water for the Poor Initiative, which will help to achieve the UN Millennium Declaration goal of cutting in half, by 2015, the proportion of people who lack safe drinking water. The United States will invest $970 million over three years, which will leverage private sources to generate more than $1.6 billion for water-related activities globally.
- The Clean Energy Initiative, which will provide millions of people with new access to efficient and clean energy services, will receive $43 million in 2003 to leverage about $400 million in investments from the United States and other governments, the private sector, and development organizations.
- The Initiative to Cut Hunger in Africa, which will focus on technology sharing for smallholders, will receive $90 million in 2003,

including some $53 million to harness science and technology for African farmers and $35 million to unleash the power of markets for smallholder agriculture.

- The Congo Basin Forest Partnership, which will promote economic development and poverty alleviation, will receive up to $53 million over the next four years to support sustainable forest management and a network of national parks and protected areas and to assist local communities, matched by contributions from international environmental organizations, host governments, G8 nations, the European Union, and the private sector.
- The reaffirmation of the commitment of President Bush to help fight HIV/AIDS, tuberculosis, and malaria through financial and technical support for the Global Fund and the International Mother and Child HIV Prevention Initiative. The Bush administration requested $1.2 billion in 2003 to combat these three diseases. These efforts will help achieve the Millennium Development Goal of halting by 2015 the spread of HIV/AIDS and the scourge of malaria and other communicable diseases.

These are all laudable initiatives, but this "new" U.S. agenda for sustainable development is disturbing to many in academia, in NGOs, and in the field.

First, it is a very different agenda from that put forward by the PCSD in its 1996 report "Sustainable America: A New Consensus." This well-respected document, which developed substantial policy recommendations and actions in both environmental justice and sustainable development, recognized the importance to the planet of a sustainable America. It devoted a full six of its seven chapters to developing sustainability initiatives in the United States, and the final chapter was about international leadership. It argued that "the U.S. must change by moving from conflict to collaboration and adopting stewardship and individual responsibility as tenets by which to live" (PCSD 1996:1). Second, in *Working for a Sustainable World: U.S. Government Initiatives to Promote Sustainable Development,* published for the WSSD by the U.S. Agency for International Development (USAID), President Bush does not even mention sustainability in America. The American lifestyle is apparently not up for debate. Instead, Bush opened by stating that "countries that live by these three broad standards—ruling justly, investing in their people, and encouraging economic freedom—will receive more *aid* from America" (USAID 2002:1). Third, the introduction to *Working for a Sustainable World* mentions that "the U.S. promotes sustainable development in 148 countries" (USAID 2002:2). Again, one assumes that among these 148 countries, the word *America* does not surface.

The Bush sustainable development agenda, if this is not an oxymoron, is simply about the greening of foreign aid: helping other countries to become more sustainable through funding partnerships, while ignoring the very compelling and comprehensive *domestic* sustainable development agenda being put forward by civil society groups in the United States. One such group, Redefining Progress, issued a direct challenge to President Bush in February 2002 through "The Johannesburg Summit 2002: A Call for Action." In it, the group was joined by major organizations such as the Natural Resources Defense Council, Friends of the Earth, the Earth Policy Institute, the Worldwatch Institute, the World Wildlife Fund, the Environmental Law Institute, Rainforest Action Network, the Nature Conservancy, Greenpeace USA, the Sierra Club and the Woods Hole Research Center. The "Call for Action," in effect a plan to close the ever-widening *sustainability gap* between knowledge and action in the United States, urged the president

> to provide an example for other nations by announcing at the Summit the specific actions you have taken and will take to reassert the importance of protecting the environment and achieving sustainable development, including to:
>
> - *ratify* and implement the major environmental treaties forged at Rio and thereafter;
> - *reduce* United States emissions of carbon dioxide and other global warming pollutants;
> - *stimulate* development and deployment of energy efficiency and renewable energy technologies;
> - *protect* critical land and marine ecosystems;
> - *provide* increased financial and diplomatic support to strengthen international environmental institutions and structures;
> - *reform* international trade and financial institutions and export credit agencies, and establish adequate safeguards to protect communities and the environment;
> - *eliminate* subsidies that cause overfishing, halt destructive fishing practices, and enforce controls on ocean pollution;

- *support* expedited development of a strategic approach to international chemicals management that is coordinated, coherent, and environmentally sound;
- *increase* U.S. assistance to developing countries to protect their environments and the global environment; and
- *defend* the fundamental democratic rights of citizens and communities around the world to protect the resources on which their lives and livelihoods depend.

This agenda was also supported by congressmen Dennis Kucinich (D-Ohio), Earl Blumenaur (D-Oregon), and George Miller (D-California), as well as by Jerry Brown, mayor of Oakland, California. At a press conference that took place before Dobriansky's press conference, the four men criticized the U.S. funding partnerships as "a recycled idea and recycled money." Their clear message was that they wished to distance themselves from Bush's foreign-aid approach to sustainable development. They argued, PCSD-style, that the United States cannot afford to withdraw from international debates on these issues. On the contrary, they argued, it is imperative that "the United States, as the richest nation and the biggest polluter, had a special obligation and opportunity." They clearly rejected the administration's position of refusal to implement the Kyoto agreements, and they argued that if the United States wants international support in its war on terrorism, then it must reciprocate by addressing the sustainable development agenda being formulated by the international community.

Jerry Brown said there were two Americas—the America of George Bush, an America of isolation, retreating from cooperation on poverty and other issues—and the America of cities and states that were aggressively pursuing their responsibilities. California, for example, passed a law that mandates a decrease in greenhouse gases produced by automobiles—the first state in the nation to do so—and Brown said that he believed many states would follow. "This America stands with the majority of the people of the United States and the people at the Summit in efforts to reduce poverty, while protecting the environment. The public are ahead of the government," said Brown. Yet nowhere in *Working for a Sustainable World* is there mention of the domestic sustainability issues raised by the PCSD in 1996 or by the congressmen, the mayor, and Redefining Progress et al. Nowhere is there mention of the urgent imperative to curb the American Dream of limitless consumption.

Some Policy Tools

While many theorists still seek "the perfect definition," there is, according to McNaghten and Urry (1998:215), a "growing impetus within the policy making community to move away from questions of principle and definition. Rather they have developed tools and approaches which can translate the goals of sustainability into specific actions, and assess whether real progress is in fact being made towards achieving them."[2] Prominent among these tools, they argue, are sustainability indicators.

Sustainability indicators help policymakers and communities understand what current conditions are (such as, What are the levels of sulfur dioxide pollution or number of people happy for their children to play in the street?), which way the indicator is going (Is there less sulfur dioxide pollution or more? Are people getting happier or less happy about their children playing in the street?), and how far we are from where we want to be (Are sulfur dioxide levels or numbers of children playing in the street nearer or further from the target values set by our community?). I agree with the need to move on to indicators as both *measures* of change and as *drivers* of change, provided that the indicators chosen reflect the breadth of sustainability concerns, not solely environmental ones, and that they are both community resonant and owned.

Some see environmental justice as a policy tool in the sustainability toolshed, but clearly others—a majority—do not. Warner (2002:37), in an Internet-based survey, found that "more than 40 percent of the largest cities (33 of 77) in the United States had sustainability projects on the web, but only five of these dealt with environmental justice on their web pages." By "dealt with" he means "makes mention of," and he categorizes the scale of "making mention" from the lowest, *education* (i.e., background information for users of the site), through *policy* (i.e., a stated policy commitment to environmental justice), to *implementation* (i.e., integration of environmental justice into sustainability). Warner continues:

> few communities were building environmental justice into local definitions of sustainability. Only five local sustainability projects made these connections: Albuquerque, New Mexico; Austin, Texas; Cleveland, Ohio; San Francisco, California; and Seattle, Washington.

This clearly illustrates the pervasiveness of the "equity deficit" I mentioned earlier. Only one city reached the implementation level, that is, integration of environmental justice into sustainability policies and plans. That city was San Francisco; the others were all at the education level, except Cleveland, which reached the policy level. Warner's conclusion is that "while environmental justice seemed to be having an impact on mainstream environmental organizations and on government agencies, this did not apparently extend to groups working on sustainability projects" (38).

It is worth looking at what San Francisco, the leading city in Warner's research, has done, or rather what it *says* it has done, in policy terms. Aside from its leadership in implementing the precautionary principle and the many just sustainability projects and programs in the Bay Area, the Sustainable San Francisco website represents the most comprehensive treatment of environmental justice as a part of a city sustainability policy. Clearly, there is a debate to be had about whether EJ should be *part of* a sustainability policy or whether EJ should infuse and inform the whole policy. Whatever the answer, and I would argue for the latter, the site did not provide as much general background on environmental justice as some others did, but there was a thorough effort to incorporate environmental justice into specific sustainability policy and to take steps toward implementation. The plan was initiated in 1993 by the San Francisco Board of Supervisors. Sustainable San Francisco organized the process of drafting a plan to include broad, long-term social goals, long-term objectives, specific actions, and community indicators for each topic area. Broader public comments were solicited in June to September 1996. The Sustainability Plan was adopted as official policy of the City and County of San Francisco in July 1997.

Environmental justice is one of fifteen topic areas of the Sustainable San Francisco Plan. Each topic area has five goals. Those for EJ are:

Sustainability Plan / Environmental Justice / Strategy

Goal 1	To establish meaningful participation in the decision-making processes that affect historically disadvantaged communities of San Francisco.
Goal 2	To create a vibrant community-based economy with jobs and career opportunities that allow all people economic self-determination and environmental health.
Goal 3	To eliminate disproportionate environmental burdens and pollution imposed on historically disadvantaged communities and communities of color.
Goal 4	To create a community with the capacity and resources for self-representation and indigenous leadership.
Goal 5	To ensure that social and economic justice are established as an integral aspect of environmental well-being and sustainability.

(Sustainable San Francisco, http://www.sustainable-city.org/Plan/Justice/strategy.htm)

The San Francisco plan also delineated three community indicators that would be used to assess progress in the area of environmental justice:

Sustainability Plan / Environmental Justice / Indicators	
Mean income level of people in historically disadvantaged communities.	↑
Proportion of environmental pollution sources in historically disadvantaged communities with respect to San Francisco's other communities.	↓
Participation of historically disadvantaged communities as a whole and their indigenous self-selected representatives in decision-making processes.	↑

An upward-pointing arrow indicates that the measurement should rise if the City is moving in the right direction; a downward-pointing arrow indicates that the measurement should fall. (Sustainable San Francisco, http://www.sustainable-city.org/Plan/Justice/indicato.htm)

Clearly, San Francisco's efforts are, in policy and aspirational terms, impressive. However, Portney (2003:171) notes that,

> although Sustainable San Francisco, a nonprofit organization, was successful in getting the city to officially adopt the plan it developed, since its adoption in 1997 it is difficult to find where in the operation of the city's governmental agencies these goals are being vigorously pursued.

Backing up Warner's research, Portney argues that "most cities that have sustainability indicators do not explicitly use social or environmental equity" (57). Yet a search of the database of the well-respected training and consulting organization Sustainable Measures[3] under *race* yields eight sustainability indicators in use in different U.S. and Canadian cities, while *ethnicity* yields two and *low income* yields fifteen.[4] What does this say about the integration of environmental justice as a tool in municipal sustainability agendas in the United States? It says that it is ad hoc at best, rather than fundamental or foundational.

Another policy tool is the Sustainability Inventory. This tool represents the second stage in a five "milestone" deliberative process developed by ICLEI (fig. 2.1). According to ICLEI (2002c:19), the inventory aims to assist municipalities in developing policies and promoting practices that maintain the integrity of natural resources over the long term while enhancing economic vitality and social well-being within a community. The Sustainability Inventory is unique because the data collected and the process of conducting an inventory provide a platform for developing holistic, collaborative management strategies within a local government. It requires data sharing among various local government departments and demands their joint collaboration in target setting and action-plan devel-

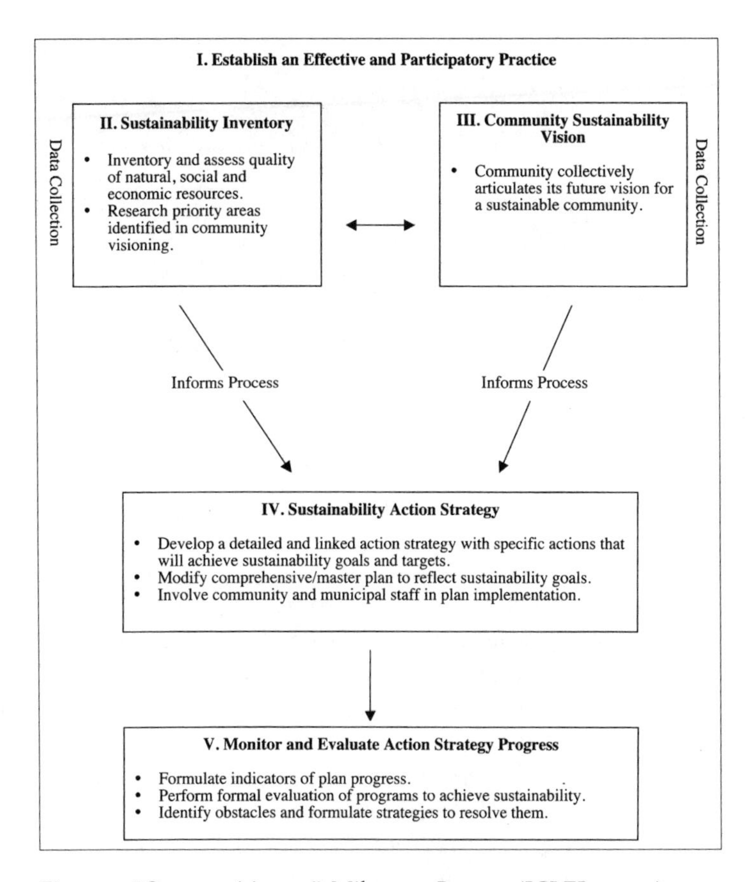

Fig. 2.1. "Communities 21" Milestone Process (ICLEI 2002c).

opment. Around the same time as the development of the inventory, municipalities should facilitate a community sustainability vision. After this, they should develop and implement a consensual sustainability action strategy with goals and targets. Finally, they should develop a means of monitoring and evaluating progress using their sustainability indicators.

Another policy tool was developed by Friends of the Earth Netherlands as part of its *Action Plan Sustainable Netherlands* (Buitenkamp et al. 1992/1993) and by Friends of the Earth Europe's *Sustainable Europe Campaign,* which is guiding national policy in the Netherlands and Denmark. This tool is the concept of *environmental space.* The term comes from the Dutch *milieugebruiks ruimte* (literally, "environmental utilization space"). The environmental space (Spangenberg 1995) is the

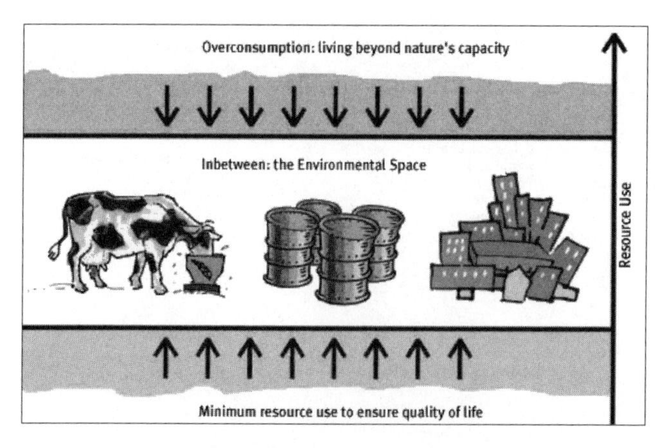

Fig. 2.2. Environmental Space (from Friends of the Earth Europe, http://www.foeeurope.org/sustainability).

sustainable rate at which we can use resources (fossil fuels, water, timber, steel, aluminum, cement, foodstuffs [land], and chlorine)[5] without precipitating irreversible damage to our ecosystems, or depriving future generations of these resources. In other words, as in fig. 2.2, it is that space in-between the minimum resource use to ensure human dignity and an acceptable quality of life (the floor) and the maximum use of Earth's capacity without depleting ecological stocks (the ceiling). This tool is like a similar tool, the *ecological footprint* (Wackernagel and Rees 1996; see note 10), only it does not aggregate resources into a single land-area-based index as the footprint does. It is based on issues of justice and equity because environmental-space targets for countries are reached by dividing the global environmental space per resource among the world's population, based on a forecast of ten billion people in 2050.[6] This will give an individual "fair share," which should then be multiplied by the forecast population of the country in 2050. This is an inter- and intragenerationally (as well as inter- and intranationally) equitable, or "fair shares," distribution of the world's resources, rather than the usual policy method of "grandfathering" of resource use, based on current, historical, or established usage of resources. For instance, Kyoto Protocol "flexible" mechanisms such as the Clean Development Mechanism and Joint Implementation assume a grandfathering approach that is unjust, since grandfathering merely allows the rich to get richer and the polluters to pollute more.

Athanasiou (n.d.) sees growing international calls for environmental justice as key in pushing the environmental-space tool. He argues that

> there are plenty of signs that a large non-rhetorical movement for global environmental justice is emerging. My claim, here, is only that this movement must find new ways to stand for equity, and that one particularly promising approach, based on asserting equal claims to limited "environmental spaces," has received too little attention.

Similarly, although more specific to just sustainability, McLaren (2003) argues that environmental space "can underpin the emerging concept of just sustainability" (22) and can "suggest how sustainable development policies can explicitly incorporate equity concerns" (34). Like Athanasiou, McLaren argues that for the environmental justice or just sustainability activist, environmental space offers "a foundation for global campaigns which can unite poor and excluded communities in North and South, rather than dividing them, and similarly unite environmental and social campaigners" (34).

This potential of the environmental-space tool to unite the "green" Northern and "brown" Southern agendas is a critical point in the building of an international agenda for just sustainability. Alas, despite the increasing usage of the *ecological footprint* concept in the United States, interest in and usage of the *environmental space* concept by U.S. policymakers wishing to guide policy, academics, or activists wishing to link with campaigners abroad has been nonexistent. I was told by one person at the Boston Foundation that this is because environmental space is about limits and is therefore a political no-go in the United States. Indeed, Buhrs (2004:431) concurs about this idea of limits, arguing that

> the importance of the notion of environmental space lies in three things: first, that it reintroduces the idea of limits at a time when, politically, the notion of limits has been pushed into the background; second, that it provides a basis for operationalizing the concept of sustainability in concrete, measurable terms; third, that it casts the notion of limits in a form that highlights distributional and equity issues.

Clearly, like the Precautionary Principle, environmental space is an area of exceptional promise for the JSP and the EJP, in the United States and worldwide.

Economic Thinking and Sustainability

One area of thinking that has been heavily influenced by sustainability, and vice versa, is economic thought (see, for instance, Daly and Cobb 1989; Jacobs 1991; Hawken 1993; Krishnan et al. 1995; Harris et al. 2001). Fundamentally, what Hempel (1999) calls the "capital theory" of sustainability says that we must not diminish our "natural capital," that is, our stocks of natural assets that yield a flow of goods for the future, such as a forest, a fish stock, or an aquifer (Hawken et al. 1999). These stocks provide important ecological services,[7] such as a harvest or flow that is potentially sustainable year after year. If the forest, fish stock, or aquifer is our natural capital, the harvest or flow is our "natural income."

There are three types of natural capital:

Renewable, such as living species and ecosystems
Replenishable, such as surface and groundwaters
Non-renewable, such as fossil fuels and minerals

Natural capital is more than just an inventory of our resources. It is the components and structures of the ecosphere, the totality of global ecosystems. Since renewable and replenishable resources are essential for life support, these forms of natural capital are fundamental and are called *critical natural capital.* They cannot be traded or allowed to depreciate.

The *strong* model of sustainability says that we should conserve and enhance our natural capital stocks and live on the income generated by them. The *weak* model says that we can lose natural capital if we substitute the equivalent "human capital."[8] For example, human-made microwave transmission and optical fibers have greatly reduced the need for natural copper. The problem with weak sustainability is that substitution does not work in many cases. For example, natural capital such as a forest is often a prerequisite for human capital such as the ability to build a sawmill. In other cases, human capital simply cannot substitute for critical natural capital.

Another economic issue with a sustainability focus is how we measure progress or development at home and abroad. Until 1991, the traditional measure was GNP (gross national product). This measure was developed in the 1940s to see how well different countries were doing in terms of wealth generation, which we then thought was synonymous with progress. GNP is a measure of human activity rewarded by a payment. The GNP of the United States is a measure of the value of goods and services produced by U.S. nationals anywhere in the world. But since 1991, the preferred measure has been GDP (gross domestic product), a measure of the value of goods and services produced within the boundaries of the United States. GDP does not, however, account for unpaid human activities like domestic and family tasks, taking care of children or the elderly, volunteer or community work, and leisure-time activities (see table 2.1). GDP is therefore only a measure of wealth, or standard of living, not of unpaid human activities that improve people's quality of life, an issue that is increasingly important in civil society and to sustainable futures. Unpaid human activities are among those that create *social capital,* the cement that keeps communities and societies functioning normally. GDP provides no information on wealth distribution, on social development, or on environmental externalities like oil spills or environmental injustices. If a country such as the United States is generating massive wealth, that is, has the world's highest GDP, but is polluting its communities massively and disproportionately by race and income and is nearing social breakdown through crime, stress, and ill health, can we say that it is progressing or developing sustainably?

Many countries are now seeking alternative ways of looking at human and physical development that take into account environmental externalities (such as the costs of pollution, depreciation of natural assets, and loss of biodiversity and ecosystems and their services) and social development (wealth distribution, housing, employment, equity, and so on). One alternative measure is the Index of Sustainable Economic Welfare (ISEW),[9] based on work by U.S. economists Herman Daly and John Cobb and developed by the University of Surrey Centre for Environmental Strategy and the New Economics Foundation in the United Kingdom. While not without its problems, such as assigning financial costs to nonfinancial impacts such as climate change and ozone depletion, the ISEW corrects GDP over a range of issues, such as income inequality, environmental damage, and depletion of environmental assets, to create an indicator that better measures how our economy delivers welfare for people.

What is the message from this "new economics"? U.S. organizations such as Redefining Progress and the Center for a New American Dream, and the New Economics Foundation in the United Kingdom (all of which have a JSI of 3), are arguing that more money (standard of living) does not mean more happiness (quality of life) (Veenhoven 1987; Seabrook 1994; Stutz and Mintzer 2003). The looming problem for Americans and for the planet is that the American Dream of a generational entitlement to a higher standard of living has become the antithesis of sustainability, with startling results. We have lost the concept of *quality* and simply replaced it with *quantity*—from *super tasting* to *super sized.* We live in built-out, sprawled-out, socio-economically and racially divided, energy- and transit-inefficient communities. We have an enormous national debt and equally sizeable personal debts. We walk less; we weigh more. We drive ever-thirstier vehicles and shop more, work more, and watch more TV, while conversing with family, friends, and neighbors less.

Unless we develop a more equitable, values-based sustainable vision for the future, guided by environmental-space targets, the materialism and individualism of today's Dream will literally end up costing us the earth. Today's 290 million Americans represent just under 5 percent of the world's population of just over six billion. We consume 30 percent of the earth's resources such as paper, plastic, energy, and chemicals and produce 25 percent of the world's major greenhouse gas, carbon dioxide, among other wastes. We Americans are leaving a huge ecological footprint[10] on the planet.

Sustainability Policies

Like Campbell (1996:301), I agree that "in the battle of big public ideas, sustainability has won: the task of the coming years is simply to work out the details and to narrow the gap between its theory and practice." While no one would claim that there is a chance of true sustainability or sustainable development in advanced industrial societies anytime soon, or that its details will be "simple," some practical policies for sustainable development that are currently being implemented by governments from local to national and by communities in different parts of the world including the United States are:

- *Eco taxes,* which shift the tax burden from good things like employment to bad things like pollution and excessive resource use. These can be very effective very quickly, as was the federal sulfur tax. The State of Minnesota has, as part of its ME3 Program (Minnesotans for an Energy Efficient Economy), a Tax and Incentives Program that "seeks to marry the goals of environmental improvement with a fair and efficient taxation system" (Sustainable Minnesota 2004). The win-win here is that research in Britain has shown that while cuts in pollution benefit everyone, they disproportionately benefit the poor (Boardman et al. 1999). Extrapolating this research to the United States, an eco-tax regimen would likely be a winner in the environmental justice movement, as well as in the sustainability movement(s).

- *Elimination of agricultural and energy subsidies,* which are environmentally damaging through their encouragement to overuse energy, fertilizer, pesticides, and irrigation water, and underuse labor in favor of mechanization. This disproportionately affects low-income, minority, and migrant laborers. Sustainable agriculture relies on recycling of nutrients, natural pest control, labor intensivity, and less artificial usage. The National Campaign for Sustainable Agriculture strap-line is "environmentally sound—economically viable—socially just—humane," reflecting their just sustainability orientation.
- *Local Exchange Trading Schemes (LETS)* enable people to decide the local unit of "currency" and trade their skills in it. "Ithaca Hours" is a local currency that pays for plumbing, renting, or even loans. In "time money schemes," the currency is the hours spent in volunteer activity, so that, for example, shopping for local elderly people becomes an alternative form of money. The Maine Time-Dollar Network of over two hundred members (from age seven to eighty-seven) is presently exchanging an average of six hundred hours of services per month. In two years, they have exchanged a total of over ten thousand hours of service with each other.
- *Affordable housing* is being financed through community finance initiatives such as community development banks, corporations, and credit unions in many cities. Organizations like ACE in Boston are moving beyond simply advocating for affordable housing toward advocating for more sustainable, energy-efficient affordable housing. Location-efficient mortgages (LEMs) in Chicago, Seattle, and San Francisco are being developed which reward certain locations (close to transit nodes). Financial institutions like LEMs because they fit community-lending criteria and help them

gain a foothold in low- to moderate-income markets. Cooperatives and co-housing options are becoming increasingly popular. The former is a proven form of multifamily homeownership. It has been used to meet a variety of needs for seniors, artists, college students, manufactured-housing park residents, and people with disabilities. Co-housing is a Danish import that builds on the increasing popularity of households in which unrelated people share a traditional house. Its distinctiveness comes from the fact that each family or household has a separate dwelling and chooses how much it wants to participate in community activities.

- *Recycling and renewable energy* are being given greater prominence in some areas. "Industrial ecology" projects such as those in Baltimore, Maryland; Brownsville, Texas; Chattanooga, Tennessee; and Port Cape Charles, Virginia, are showing how industrial systems can be made to mimic the closed-cycle patterns of natural systems. Such systems, often in "Industrial Eco-Parks," are different from the traditional model of industrial activity because the flow (including cycles) and stock of materials and energy is optimized. Emphasis is placed on efficiency; waste reuse, recovery, and exchange; and the minimization of adverse environmental impact, toward a goal of zero waste.
- *Efficient transportation systems* are being developed which replace energy-intensive automobile transport with high-speed trains (such as the Acela, from Boston to Washington, D.C.), public transit (such as the San Diego Trolley), and greater use of bikes and walking (as in Cambridge, Massachusetts). City and suburban redesign through "smart growth" and "New Urbanism" projects minimizes transportation needs through mixed-use developments. This and other transportation initiatives create a focus on *access* rather than *mobility,* which promotes equity-based thinking. The promotion of more-equitable systems of transportation is an area of EJ activism in which "cooperative endeavors" (Schlossberg 1999) are taking place between EJ groups and groups representative of the JSP. In Boston, a most welcome development is the partnership between the Washington Street Corridor Coalition (WSCC), an EJ group looking to push the Massachusetts Bay Transportation Authority (MBTA) into using light rail on the Washington Street corridor of the MBTA's Silver Line bus service, and the Boston Group

of the Sierra Club (a group representing the NEP), which is expanded upon in chapter 5.
- *Community Supported Agriculture* schemes (such as Appleton Farms CSA, in Ipswich, Massachusetts, or Community Farms in Europe or Food Guilds in Japan) and farmer's markets (such as Somerville Farmer's Market, in Somerville, Massachusetts) are becoming increasingly popular in U.S. cities. In some inner urban areas, according to Ashman et al. (1993), "community food security" has emerged as an issue in needs-assessment surveys by the environmental justice groups the Concerned Citizens of South Central Los Angeles (CCSLA) and the Mothers of East Los Angeles (MELA).

Note the breadth of concern of this limited selection of sustainability policies. They are not only environmental policies but focus on justice and equity as well. Following Campbell's (1996) point, the question now becomes, Can we achieve sustainable development and sustainable communities by tweaking existing policies, which we are doing at present, or do we need a rethink, a paradigm shift? The policy areas listed above (and those listed in table 2.1) are valuable demonstrations of what we could achieve and, in certain cases and localities, are achieving, but they are still "best practices," not "ordinary" or "mainstreamed" practices. And, if we need a paradigm shift, will the NEP alone deliver sustainable communities, when, as Taylor (2000) has argued, it is virtually devoid of an appreciation of social justice? Both the NEP and EJP are deficient in prescribing and delivering sustainable communities in some ways (see table 3.1). However, the emergent JSP, as a bridge between the two, does not have these deficiencies and is predicated on delivering sustainable communities, as I show in chapter 3.

Characteristics of a Sustainable Community

The move away from the reductionist *environmental* sustainability toward a more joined-up acceptance of the environmental, social, and economic principles of sustainability can be seen most clearly in the movement toward sustainable communities. Unlike the contested nature of the concepts of sustainability and sustainable development, the concept of sustainable communities seems to be less problematic. For this reason, it is often the preferred focus of discussion. Similarly, despite what I earlier called "institutional learning in reverse" to describe the Bush administration's attitude toward sustainable development, the move toward sustainable communities in the United States can be seen at both the theoretical level and the practitioner level (Kline 1995; Campbell 1996; Beatley and Manning 1997; PCSD 1996, 1997; Roseland 1997, 1998; Hempel 1999; Mazmanian and Kraft 1999; Shutkin 2000; Agyeman and Angus 2003; Portney 2003), through smart-growth programs, individual municipality programs,[11] and programs such as the ICLEI's Communities 21 Program (ICLEI 2002c).

According to Hempel (1999), the concept and protagonists of sustainable communities (and sustainability more generally), like those for environmental justice (see Cole and Foster 2001 and Faber and McCarthy 2003), come from many roots: the *natural capital/capital theory* approach of economists, the *urban design* approach of land-use planners and architects, the *ecosystem management* approach of ecologists and resource managers, and the *metropolitan governance* approach of regional policy planners. He notes, however, that "none of these are mutually exclusive, nor can they be said to constitute forms of identity that are commonly or self-consciously accepted by participants in the sustainable communities movement" (1999:53). I mentioned earlier that the main drivers of the sustainability project are environmentalists. Each of Hempel's roots focuses largely on *environmental* sustainability, except the metropolitan governance approach, which has a more broad "politically oriented conception of sustainable development" (58) (see, for instance, ICLEI 2002c; Orfield 1997; Delgado 1999).

Similarly, the concept is not a rigid one-size-fits-all formulation; it does not imply a set design for a particular type or location of town or city. Rather, "a sustainable community is continually adjusting to meet the social and economic needs of its residents while preserving the environment's ability to support it" (Roseland 1998:14). The sustainable communities movement, according to Hempel (1999:51), may be seen as a

> reaction to decades of frustration felt by transportation and land-use planners, municipal officials, neighborhood activists, downtown business leaders, and environmental groups faced with their inability to constrain and effectively manage urban sprawl and its accompanying social and environmental ills.

TABLE 2.1
Characteristics of a Sustainable Community

A Sustainable Community Seeks to:	
Protect and enhance the environment	• Use energy, water, and other natural resources efficiently and with care • Minimize waste, then re-use or recover it through recycling, composting, or energy recovery, and finally sustainably dispose of what is left • Limit pollution to levels that do not damage natural systems • Value and protect the diversity of nature
Meet social needs	• Create or enhance places, spaces, and buildings that work well, wear well, and look well • Make settlements "human" in scale or form • Value and protect diversity and local distinctiveness and strengthen local community and cultural identity • Protect human health and amenity through safe, clean, pleasant environments • Emphasize health service prevention action as well as cure • Ensure access to good food, water, housing, and fuel at reasonable cost • Meet local needs locally wherever possible • Maximize everyone's access to the skills and knowledge needed to play a full part in society • Empower all sections of the community to participate in decision-making and consider the social and community impacts of decisions
Promote economic success	• Create a vibrant local economy that gives access to satisfying and rewarding work without damaging the local, national, or global environment • Value unpaid work • Encourage necessary access to facilities, services, goods, and other people in ways which make less use of the car and minimize impacts on the environment • Make opportunities for culture, leisure, and recreation readily available to all

Source: DETR (1998).

Hempel (1999:48) describes a sustainable community as one in which:

> economic vitality, ecological integrity, civic democracy, and social well-being are linked in complementary fashion, thereby fostering a high quality of life and a strong sense of reciprocal obligation among its members.

Table 2.1, developed by the Local Government Management Board in Great Britain in 1994, presents the characteristics of an ideal sustainable community that espouses these environmental, social, and economic goals.[12] These characteristics and goals are remarkably similar to the Elements of a Sustainable Community that were developed by the board of the Institute for Sustainable Communities (ISC) in Vermont and subsequently utilized by the PCSD (1997) in its Task Force Report, "Sustainable Communities."[13] This positive news indicates that, among nations of the North, caveats about "green" and "brown" agendas notwithstanding, there is some agreement about where we want to be with regard to sustainable communities. As Campbell (1996:301) said, now we need to "work out the details, and to narrow the gap between . . . theory and practice."

The seventeen individual characteristics listed in table 2.1 are not a simplistic wish list of unachievable goals. They are all scientifically, technically, economically, and socially feasible, and examples of communities practicing one or two of them can be found in Best Practices databases around the world.[14] A growing number of U.S. initiatives are also demonstrating some of these characteristics, including in Arcosanti, Arizona; Civano, Arizona; Crestone/Baca, Colorado; Dancing Rabbit Eco-village, Missouri; Davis City, California; Dreamtime Village, Wisconsin; Earthaven Eco-village, North Carolina; EcoCity, Cleveland, Ohio; The Farm, Tennessee; Ithaca, New York; LA Eco-village, California; and Sirius, Massachusetts.

Note that the majority of these U.S. projects are in rural areas and in small towns. No major U.S. city with a population above two hundred thousand has declared, complete with goals, timelines, and targets, that it will become a sustainable community at some prescribed point in the future, although many cities are, according to Portney (2003), "taking sustainability seriously." The point is that while new communities or small communities reoriented toward sustainability are beneficial and can demonstrate the logic of sustainability, and the implementation of sustainability policies and plans, the major challenge is how to make our (and the world's) larger cities more sustainable. The programs and projects from U.S. cities described in chapter 4 are a small attempt to do just that.

However, Portney (2003:31) asks not whether cities have become more sustainable, a question that he sees as "premature and perhaps presumptuous," but how seriously cities are taking sustainability. The focus of his analysis is on "sustainability initiatives," "any set of activities, programs, policies, or other efforts whose purpose is explicitly to

TABLE 2.2
Correlations between the Index of Taking Sustainability Seriously and Population and Employment Characteristics in Twenty-Four Cities

Independent Variable	Pearson Correlation Coefficient	Significance
Percent African American, 2000	−.391	0.06
Percent African American, 1990	−.340	0.10
Percent Hispanic, 2000	−.278	0.18
Percent Hispanic, 1990	−.241	0.26
Percent under 18 years old, 1990	−.495	0.01
Percent over 65 years old, 1990	.147	0.50
Median age of the population, 1990	.621	0.00
Percent high school graduate, 1990	.501	0.01
Percent employed in manufacturing, 1990	−.547	0.01
Percent employed in service sector, 1990	.094	0.66

Source: Portney (2003).

contribute to becoming more sustainable" (35), and more specifically on "sustainability plans," "perhaps the single most important element in assessing the seriousness of a city's efforts toward achieving sustainability" (37). Using data from cities' initiatives and plans, Portney develops an Index of Taking Sustainable Cities Seriously out of thirty-four specific activities[15] (e.g., car pool lanes, indicators project, green building programs) in seven categories (e.g., pollution prevention and reduction efforts). Each activity is equally weighted (which Portney acknowledges is a problem: is a program to promote bicycle riding more or less worthy than renewable energy use by city government?), giving for the twenty-four cities in his study an index from 0 to 34, or, in practice, from 6 (Milwaukee) through 14 (Cambridge, Massachusetts) to 30 (Seattle).

Portney then uses the index as a dependent variable to investigate some basic correlates. Independent variables such as population size, rapid economic growth, median family income, median household value, total and per capita local government spending, and unemployment rate were either weakly or not correlated with his Index of Taking Sustainable Cities Seriously. However, his significant correlates are shown in table 2.2. They tended to be

> demographic characteristics—median population age, the percentage of the population below 18 years of age, the percentage of high school

> graduates, and the percentage of African Americans residing in the city —and one indicator of the local economy—the percentage of the labor force employed in manufacturing industry. (2003:238)

He concludes that

> what this probably means is that the cities that need sustainability most —cities that are reliant on relatively more polluting manufacturing industries as the base of employment, and cities with younger populations—are the cities that tend to take sustainability less seriously. (238)

In other words, cities that do take sustainability seriously are less-polluted cities in the first place.

Given the nature of my arguments throughout this book, it is worth dwelling here on cities with more polluting industry and higher numbers of African Americans. These are likely to be cities that, in Massachusetts's terminology from chapter 1, have large numbers of Environmental Justice Populations and are precisely those cities that *should* be taking sustainability more seriously. In explaining his data on African Americans, Portney says that "whether it is because these cities face what they consider to be more pressing problems or because members of these populations simply do not place a high value on trying to achieve sustainability is impossible to say here" (2003:234). There are two issues here. First, African Americans, as Mohai (2003) has shown, do place a high value on environmental sustainability. They are as concerned as whites about global environmental and nature preservation issues but are *more* concerned than whites about pollution issues, especially in local neighborhoods. This heightened concern is attributed by Mohai and Bryant (1998) to the "environmental deprivation" effect: the result of African Americans living in more-polluted neighborhoods. Second, given the dominance of the *environmental* discourse of sustainability (and the *environmental* dominance of Portney's 2003 Index, on which more later), most African Americans will see environmental sustainability in the same way as the environmental justice movement saw traditional environmentalism twenty years ago: important, but in need of a clear linkage to the doorstep issues of housing, jobs, and racism. Taken together, these points support arguments surrounding the need for coalitions between EJP and JSP activists to get involved in municipal sustainability initiatives, which, as Warner (2002) has shown, are sadly lacking in equity and justice perspectives.

One problem with Portney's index, which may help explain the African American data, is that it is primarily an *environmental-economic* index. Out of the thirty-four specific activities in seven categories mentioned above, the index lacks a social or social-justice category. Certainly, if Portney's methodology were transferred from developing the index for cities in the North to cities in the South, where the antipoverty "brown," as opposed to the environmentalist "green," sustainability agenda dominates, then a social-justice category would be created as a matter of course.

It will be noted, however, that if a robust sustainability inventory were completed, which could assess whether the characteristics of a sustainable community (as shown in table 2.2) were present in any given community, and where the imprimatur *sustainable community* meant doing not one but *all* these tasks together, then no community on earth would be classified as a sustainable community. The point is that these characteristics, which, among others, emphasize environmental protection for all, valuing diversity and cultural identity and creating rewarding work, are for stakeholders in communities, NGOs, and policymakers in local governments to use as guidance, as a framework in the move toward more-sustainable communities, perhaps using the "five milestone" methodology suggested by ICLEI in fig. 2.1. This synergistic approach, argues Roseland (1998:2), "will enable our communities to be cleaner, healthier, and less expensive; to have greater accessibility and cohesion; and to be more self-reliant in energy, food and economic security than they are now."

While the elements of a sustainable community include managing and balancing environmental, economic, and equity concerns, in order for these elements to be fully realized, two further critical elements are needed: democracy and accountability.

First, democracy is needed to coordinate and balance the process. Indeed, Roseland (1998:24) notes,

> for people to prosper anywhere they must participate as competent citizens in the decisions and processes that affect their lives. Sustainable development is thus about the quantity and quality of empowerment and participation of people. Sustainable development therefore

> requires mobilizing citizens and their governments toward sustainable communities.

A process of deliberative, democratic civic renewal and enhanced civic engagement is seen by many, including myself, as essential to the process of developing sustainable communities. Indeed, Selman and Parker (1997:172) note that sustainable development cannot become a self-sustaining process if it depends solely on statutory, expert-led intervention on a centrally mandated set of priorities and that

> this has major consequences for those charged with facilitating local sustainability strategies. Its implications include the ways in which local policy is debated and implemented, as well as the wider focus of environmental and community change, and the ways in which local government and stakeholders can work together to improve local quality of life.

As a challenge to the dominant U.S. political democratic model of representative democracy, deliberative approaches such as DIPS aim to increase opportunities for civic engagement, where that civic engagement forms an ongoing critical dialogue regarding alternatives or improvements to contemporary political authority (Smith 2003). Indeed, the rise of local sustainability initiatives and LA21 processes since Rio in 1992 has been matched by a concomitant increase in the use of DIPS. It would be imprudent to say there was a direct positive correlation, but it seems to me and others (Holmes and Scoones 2000) that the linkages between the two are no coincidence. LA21, according to Freeman et al. (1996:65), requires the development of "fresh and innovative methods of working with and for the community," and DIPS such as mediation and stakeholder group engagements, citizen's juries, community visioning, citizen initiative and referendums, deliberative polling, and study circles are just such methods.

According to Study Circles Resource Center,

> a study circle is a group of 8–12 people from different backgrounds and viewpoints who meet several times to talk about an issue. In a study circle, everyone has an equal voice, and people try to understand each other's views. They do not have to agree with each other. The idea is to share concerns and look for ways to make things better. A facilitator

> helps the group focus on different views and makes sure the discussion goes well. (Study Circles Resource Center 2004)

In New Hampshire, Portsmouth Listens used the study circle technique. Portsmouth Listens is a collaborative formed by the city, the Citywide Neighborhoods Committee, the Greater Portsmouth Chamber of Commerce, and residents. The aim was to get as wide an input into the city's new ten-year master plan as possible. The plan looks at a breadth of topics integral to city life, including land use; housing; economic development; transportation; recreation; cultural, historical, and artistic resources; natural resources; and sustaining the vitality of the downtown core.

A second, and related, essential element of a sustainable community is accountability. Foster (2003:803) argues that

> accountability should be concerned with how two factors manifest themselves in development decision-making processes. The first factor concerns the quality of participation among members of a community affected by development decisions. The second factor concerns the representativeness of individual participation within those communities.

She concludes that "accountability can be realized in collaborative decision-making structures built on deliberation, community representation, and equitable participation" (2003:803).

Popatchuk (2002) sees a broader movement toward what he calls "collaborative communities,"

> helping complex urban policy and decision-making systems work smarter and remain grounded in the will of citizens through the use of increasingly sophisticated deliberation, collaboration, consensus building and conflict resolution tools. (3)

While Popatchuk sees the Healthy Communities movement as the most notable exponent of collaboration, he notes that "this family of efforts also includes the Sustainable, Livable and Safe Communities movements" (10). Collaborative and deliberative tools mentioned by both Foster (2003) and Popatchuk (2002) are characteristic of the JSP and are very accessible and immediately usable as a method of increasing the interest in and diversity of meetings.

Civic Environmentalism and Sustainable Communities

Brulle's (2000) nine discourses of the U.S. environmental movement mentioned in the introduction are the discursive frames of EMOs. Yet, over the past ten years, civic environmentalism has emerged from within administrative institutions as the dominant discursive frame on sub-national environmental policymaking.[16] DeWitt John, formerly of the National Academy of Public Administration, was the first person to articulate and name civic environmentalism as an emergent policy framework and discursive frame that recognized the limits of top-down environmental regulation, during the "regulatory reinvention" initiative of the Clinton-Gore years. The approach stems from an increasing awareness, in the EPA and elsewhere, that centrally imposed, media-specific environmental policy found in legislation like the Clean Air Act and the Clean Water Act is not sufficient for dealing with contemporary environmental problems and that more flexible and collaborative solutions should be found.

Since John's (1994) work, there have been various interpretations of the concept of civic environmentalism by authors such as the EPA (1997); Roseland (1998); Sabel et al. (1999); Friedland and Sirianni (1995); Landy et al. (1999); Hempel (1999); Mazmanian and Kraft (1999); and Shutkin (2000). Shutkin (2000), who was co-founder of ACE (see chapter 5), like Roseland (1998), Hempel (1999), and Mazmanian and Kraft (1999), interprets civic environmentalism more broadly than his contemporaries (cf. Layzer 2002).[17] To these researchers, civic environmentalism is the idea that members of a particular geographic and political community

> should engage in planning and organizing activities to ensure a future that is environmentally healthy and economically and socially vibrant at the local and regional levels. It is based on the notion that environmental quality and economic and social health are mutually constitutive. (Shutkin 2000:14)

In addition, these researchers see a civic environmentalism that stresses a commitment to community organizing and promotes a healthy skepticism about the promise of science and expertise to solve social ills by themselves. Table 2.3 makes a distinction between these discourses of civic environmentalism. Agyeman and Angus (2003) have termed them

TABLE 2.3

Narrow-Focus and Broad-Focus Civic Environmentalism

	Narrow-Focus Civic Environmentalism	Broad-Focus Civic Environmentalism
Main contributors	John (1994), EPA (1997), Sabel et al. (1999), Friedland and Sirianni (1995), Landy et al. (1999)	Shutkin (2000), Roseland (1998), Hempel (1999), Mazmanian and Kraft (1999), Foster (2002)
Central premise	Stresses limits of top-down command and control environmental regulation. Civic environmentalist policies are best suited to dealing with the local nature of contemporary environmental problems.	Stresses interdependent nature of environmental, social, political, and economic problems. Civic environmentalism is fundamentally about ensuring the quality and sustainability of communities.
Central focus	The focus is on the interconnected nature of environmental problems. Using an ecosystem focus, the argument is that environmental problems do not correspond to political boundaries.	The focus is on the connections between environmental, economic, and social issues such as urban disinvestment, racial segregation, unemployment, and civic disengagement.
Contribution to sustainable communities	Can only help achieve the environmental goals of a sustainable community, namely, to "protect and enhance the envirnoment," e.g., pollution control and protection of biodiversity. (see table 2.1).	Can help to "protect and enhance the environment" while "meeting social needs" and "promoting economic success," i.e., meets all the goals of a sustainable community (see table 2.1).
Nature of change	Technical, reformist. Policy change to incorporate community perspectives.	Political, transformative. Change requires paradigm shift.
On the role of the citizen	Passive citizenship; focus on rights of citizen access to legislative and judicial procedures, community right-to-know laws.	Active citizenship; focus on responsibilities of the citizen to the environmental, social, and economic health of the community.
Role of social capital	Builds social capital as citizens gain access to the regulatory and public interest arena. But precludes broader conception of and growth of social capital because of unrepresentative nature of local environmental action.	Environmental, economic, and social decline mirrors decline of social capital. Increasing social capital and networks of social capital is essential for developing sustainable communities.
Stance on environmental justice	Environmental injustice is mostly related to lack of access to, and protection from, public policy. The primary focus is on procedural justice.	Environmental injustice is a result and cause of social, economic and racial inequity. The focus is on both procedural and substantive justice.

narrow-focus and *broad-focus* civic environmentalism. As with any general classification, there are bound to be interpretive discrepancies (see note 18, for example). The categorization is based on a content analysis of a range of key civic environmentalist texts and some examples, ranging from those with regional collaboration and oversight by federal agencies (see, for instance, Landy et al. 1999; Wondollek and Yaffee 2000) to more grassroots, bottom-up projects initiated by communities with little or no formal agency oversight (Shutkin 2000).

Narrow-Focus Civic Environmentalism

Narrow-focus civic environmentalism emerged with the seminal work of John (1994), followed by the work of Landy et al. (1999), Sabel et al. (1999),[18] and Friedland and Sirianni (1995). It recognizes that the geography of environmental problems rarely meshes with existing political boundaries (Landy et al. 1999:3), and it is a clear departure from the first generation of national environmental policies that tended to impose top-down and prescriptive solutions to address one problem at a time, independent of the circumstances in a particular place (ibid.). The approach plays a large role in what has been termed new or *second-generation environmentalism* by groups such as the Reason Public Policy Institute.[19]

The EPA's response to the emergence of civic environmentalist strategies has been the development of Community-Based Environmental Protection (CBEP) (EPA 1997). CBEP is an example of "devolved collaboration," described by Foster (2002:460) as "closely associated with the emergence of alternative dispute resolution techniques in environmental decision making." John and Mlay (1999:361) write that "if civic environmentalism is essentially an ad hoc bottom-up process of local problem solving and decision making, community-based environmental protection is top-down support for bottom-up initiatives." However, they continue by arguing that CBEP projects have had "limited success in igniting a civic process, and often the programs simply bring together the usual list of agency staff and advocates to write a plan that leads to greater understanding of an issue but little or no action" (1999:363). Foster (2002:463) sees this as a result of devolved collaboration reproducing "many of the deeper and troublesome aspects of current decision-making processes" such as distributional inequities resulting from racial and class issues, which she argues has the effect of "further entrenching them in the landscape of environmental decisions."

John and Mlay's admission that CBEP has "had limited success in igniting a civic process" reflects a fundamental problem with interest-group pluralism. CBEP has, in common with much traditional or reform environmentalism, merely rounded up the usual suspects—representatives of narrow-focus civic environmental groups and adherents of the NEP—rather than creating an inclusive, representative, and deliberative civic process. Foster (2002:498) sees this failure as the lack of "identification of a core set of normative goals—including procedural and distributional justice—for environmental and natural resource decision making [that] can be useful in both increasing meaningful participation by local actors and communities and strengthening the hands of central authorities to ensure these goals are met." Indeed, it seems that the civic *process* of participation has been forgotten in the single-minded pursuit of the environmental *product.* In this respect, CBEP is, as its name suggests, more *environmental* than *civic.*

As is outlined in table 2.3, the narrow-focus discourse of civic environmentalism, exemplified here by CBEP, tends to stress reform and policy change to incorporate community perspectives. In this way, the narrow-focus orientation is still very municipally focused, with limited delegation of power. Narrow-focus civic environmentalism also stresses the pluralistic rights of the citizen to information and to access the policy arena, despite Foster's (2002:470) warning that "political and economic power and technical ability are the currencies of pluralism."

A Narrow-Focus Example: The Chesapeake Bay Program

An example from within the narrow-focus discourse would be the Chesapeake Bay Program, perhaps the most often cited case of a collaborative, devolved approach to restoring a severely damaged watershed. Here, the EPA, the states of Maryland, Virginia, and Pennsylvania, and Washington, D.C., together with the Chesapeake Bay Commission, used voluntary measures such as education and technical assistance to achieve the program's goals. Since its inception in 1983, the Bay

Program's highest priority has been the restoration of the bay's living resources—its finfish, shellfish, bay grasses, and other aquatic life and wildlife. Improvements include fisheries and habitat restoration, recovery of bay grasses, nutrient and toxic reductions, and significant advances in estuarine science. Other cases of narrow-focus civic environmentalism include the Everglades restoration (John 1994) and the Great Lakes Basin (Rabe 1999).

Broad-Focus Civic Environmentalism

Broad-focus civic environmentalism, which has not been central to the EPA's CBEP, goes beyond the place-based environmental problem solving of more narrow-focus approaches. Broad-focus civic environmentalism also stresses the local and complex nature of environmental problems. Critically, though, it represents a greater appreciation of the holistic nature of sustainability, a greater commitment to the sharing of power between local governments and communities, and a greater level of empowerment among citizens. In this sense, it could be argued that the narrow-focus orientation is based on the lower rungs of Arnstein's (1969) "ladder of citizen participation," which signify limited citizen power, whereas the broad-focus orientation typifies the upper rungs, which are characterized by more power sharing.

The discourse of broad-focus civic environmentalism offers a set of emerging policy initiatives that variously

> seek to restore a sense of place, mixed-use development, and environmental quality that are at the core of healthy [sustainable] communities and look to social policy solutions that address the connections between environmental problems and economic and social issues such as urban disinvestment, racial segregation, unemployment, and civic disengagement. (Shutkin 2000:12)

Clearly the discourse and interpretations of broad-focus civic environmentalism, with its spotlight on urban disinvestment, racial segregation, unemployment, and civic engagement, together with a vision of political transformation and paradigm shift, is the de facto discourse of just sustainability.

A Broad-Focus Example: Dudley Street Neighborhood Initiative

An example of the broad-focus discourse is one of the classic cases of urban community revitalization: the development of Dudley Street, which straddles the Roxbury-Dorchester line in Boston, by the Dudley Street Neighborhood Initiative (DSNI). DSNI, a coalition partner of ACE, is an excellent example of what can happen when community leaders change the framing of their activism from reactivity to proactivity, from seeing community deficits to seeing community assets.

Medoff and Sklar (1994), who chronicled the DSNI effort in their book *Streets of Hope,* call this kind of activism "holistic development": a combination of human, economic, and environmental development. Similarly, the DSNI itself produced a report in 1993 called "Framework for a Human Development Agenda." At that time, the group did not use the terms *sustainable development* or *sustainable community,* although this is precisely what it was, and is still, building. In Pitcoff's (1999) interview with Greg Watson, who was at that time executive director of DSNI, Watson did not use the term *sustainable community,* but he was clearly talking about just that:

> when we did our urban village visioning process and talked about our economic power strategy, what became clear was that economic power was not the end, it was a means towards quality of life. That's really what people want for where they live. (Pitcoff 1999:23)

More recently, DSNI has developed a Sustainable Development Committee, which, among other things, assesses commercial and housing proposals. In order to do this, the committee developed what it calls a "Community Impact Assessment Tool," which measures ownership, representation, environmental soundness, and economic feasibility.

DSNI's board is diverse, reflecting the makeup of the local community. It works to implement resident-driven plans with partners including Community Development Corporations (CDCs), other nonprofit organizations and religious institutions serving the neighborhood, banks, government agencies, businesses, and foundations. DSNI's approach to neighborhood revitalization is comprehensive (physical, environmental, economic, and human). The group was formed in 1984 when residents of the Dudley Street area came together out of fear and anger to revive their neighborhood, which was nearly devastated by arson, disinvestment, neglect, and redlining practices, and protect it from outside speculators. DSNI is the only community-based nonprofit in the country that has been granted eminent-domain authority over abandoned land within its boundaries.

DSNI's vision was to create an "urban village" with mixed-rate housing. However, it soon realized that retaining community-driven development would not be sufficient to halt the kind of gentrification that now displaces residents in other parts of the city. DSNI's solution was the creation of a community land trust, Dudley Neighbors, Inc., which uses a ninety-nine-year ground lease that restricts resale prices to keep the land available for affordable housing. Out of the 500 units built by DSNI, about 170 belong to Dudley Neighbors, Inc.

Narrow-Broad Differences

Having given examples of both narrow- and broad-focus civic environmentalism, I agree with Layzer (2002) that most case studies of civic environmentalism to date have been used to illustrate but not evaluate the effectiveness of each approach. Layzer (2002:6) writes, "there has been little systematic investigation of how to translate the various approaches of civic environmentalism into durable programs that actually protect ecosystems."

My distinction between the narrow-focus and broad-focus discourses of civic environmentalism, which parallels the "biocentric" and "anthropocentric" environmentalisms of Gould et al. (2004), does not mean to suggest that narrowly focused environmental discourse or action is somehow less valuable than that which is broadly focused. What I am suggesting, however, is that it will be far more difficult to achieve what Hempel (1999:48) describes as the "economic vitality, ecological integrity, civic democracy, and social well-being" that are necessary for the development of sustainable communities (as outlined in table 2.1 and developed by ICLEI in its Communities 21 Program, ICLEI 2002c) without a more broadly based, social, economic, and political analysis. Foster (2002:498) makes a similar point: "this evolving approach [devolved collaboration] is indifferent to the ecological, social and political conditions necessary to realize its own promise." Brulle (2000:191), too, is of the opinion that reform environmentalism's "*scientific* analysis of environmental problems has developed a strong critique of the ecological effects of our current institutional structure. However, this alone is not sufficient to develop an alternative vision" (my emphasis).

This promise, this alternative vision, I would argue, can only be provided by broad-focus civic environmentalism. This point becomes obvious when one sees that narrow-focus civic environmentalism only looks at the interconnected nature of environmental problems themselves, whereas the broad focus is on the connections between environmental problems and economic and social issues such as urban disinvestment, racial segregation, unemployment, and civic disengagement (see table 2.3). Based on an argument made by Shutkin (2000), I see broad-focus civic environmentalism as more about the *civic* aspects of civic environmentalism than the *environmentalism* aspects, which I see as more the focus of narrow-focus civic environmentalism.

The interpretive approach would say that in terms of experience, core values and beliefs, political ideology, and environmental philosophy, broad-focus civic environmentalism, like its soulmates environmental justice and just sustainability, has grown out of a more diverse political and activist base than narrow-focus civic environmentalism, which comes firmly from the traditional environmental movement and the NEP. In this sense, broad-focus civic environmentalism, again like just sustainability, could be called *third-generation environmentalism*. Organizations such as Boston's DSNI and ACE, or San Francisco's Urban Habitat Program (UHP), or Chicago's Center for Neighborhood Technologies (CNT), with their broader, proactive sustainable community agenda and a strong environmental justice and just sustainability focus, contrast with the ecosystem-based projects, such as the Chesapeake Bay Program, that characterize much of narrow-focus and CBEP civic environmentalism.

Conclusion

Through the discourse and interpretation of broad-focus civic environmentalism—and not the narrow-focus type that currently dominates sub-national environmental policymaking, the NEP, and much of the traditional/reform environmental movement—a space has opened for "cooperative endeavors" (Schlosberg 1999), coalition building between the environmental justice and just sustainability movements. This nexus represents what Evans and Boyte (1986:xix) call "free spaces": "settings which create new opportunities for self definition, for the development of public and leadership skills, for a new confidence in the possibilities of participation, and for wider mappings of the connections between movement members and other groups and institutions." Whether this free space becomes big enough, and safe enough, for movement fusion (Cole and Foster 2001) between the environmental justice and just sustainability movements is another matter.

3

Just Sustainability in Theory

Having given a background to the discourses of environmental justice and sustainability, sustainable development and sustainable communities, to narrow-focus and broad-focus civic environmentalism and some linkages between them, it is my intent in this chapter to more fully map the nexus between the discourses of environmental justice and sustainability.[1] In other words, I now want to characterize the JSP as a bridge between the NEP and the EJP, noting its similarities to and differences from each. I show its global and local relevance and its four main focal areas of concern: quality of life, present and future generations, justice and equity, and living within ecosystem limits.

My caveat, mentioned in the introduction, is that my characterization largely surrounds two major players in the JSP: the environmental justice movement and organizations using the just sustainability discourse. Other than passing references to other inhabitants, such as peace, indigenous, spiritual, women's, civil rights, labor, and antiracist groups, I do not intend to fully explore their roles here.

Mind the Gap: Some Reasons for the NEP-EJP Divide

Before I characterize the JSP, I want to briefly investigate the reasons for the gap between the NEP and the EJP in terms of demographics, discourses, and movement-building practices, because if we are to occupy the free space, develop more cooperative endeavors, and ultimately look toward movement fusion, then we need to know why the gap existed in the first place.

First and foremost, as I mentioned in the introduction, the two movements came from very different places, resulting in different approaches, tactical repertoires, and languages/vocabularies. The environmental justice movement can be understood as a popular, community, grassroots, or bottom-up reaction to external threats, while the sustainability agenda, then movement, emerged in large part from expert international processes and committees, governmental structures, think tanks, and international NGO networks. In this sense, sustainability as a policy approach can be understood as a more exclusive or top-down phenomenon. Paradoxically, however, as we saw in the previous chapter, the majority of current sustainability action is generally seen as being through local action involving multistakeholder partnerships (ICLEI 2002a).

Second, as in the relationship between labor and mainstream environmentalism (Gould et al. 2004), there has been a history of mistrust between the environmental justice movement and the sustainability movement's precursor, the environmental movement. I will not revisit the blow-by-blow account here, as other authors have done so more completely than space allows (Shabecoff 1990; D. Taylor 1992, 2000; Gottlieb 1993; Bullard 1993, 1994). Suffice it to say that there were two key complaints or demands that EJ activists made, according to Schlosberg (1999:9):

> while the lack of minority representation in the offices and on the boards of major environmental groups was a focus, a more telling complaint centered on the movement's focus on natural resources, wilderness, endangered species and the like, rather than toxics, public health and the unjust distribution of environmental risk.

Related to this is the fact that Big 10 organizations tend to be very hierarchical, centralized, Washington based, and therefore distant from community concerns.

Third, and again as in the relationship between labor and mainstream environmentalism (Gould et al. 2004), the issues of class, social location, and demography cannot be ignored. There are many studies in the United States (Devall 1970; Milbrath 1984; ECO 1992; Lichterman 1995) and in the United Kingdom (Cotgrove and Duff 1980; Lowe and Goyder 1983; Gaines and Micklewright 1988; Agyeman 1990) that show that traditional environmentalists, and the organizations that hire them, are predominantly middle and upper-middle class, male, and white. Environmentalists not working in the EJ movement tend to have a college or postgraduate degree, work in a professional job, and own a home. Contrast this with Taylor's study of 331 EJ organizations, which showed that

> the leadership . . . is shared almost equally between males and females. Fifty percent. . . . have female presidents or chairs. . . . Twenty-three percent describe themselves as having a predominantly African American membership, 11% are predominantly Latino, 1% are predominantly Asian, 27% are predominantly Native American, another 27% are composed of a mixture of people of color groups, and 12% are a mixture of people of color and whites. (Taylor 2000:533)

The EJ movement, it could be argued, because of its popular as opposed to expert origins, is far more diverse in all ways than is the sustainability movement.

Fourth, there is reluctance among some EJ activists, as I show in more detail in my case study of ACE in chapter 5, to engage in what is perceived as a white, middle-class discourse—namely, sustainability or, more correctly, *environmental* sustainability.

Finally, and related, another set of differences relates to interpretations, reference points, terminologies, and individualistic versus communitarian approaches to movement building. To a certain extent, both environmental justice and sustainability activists (especially those in organizations using the discourse and interpretations of the JSP) cover similar issues, see similar problems such as racism, classism, inequity, and disinvestment and share injustice framings. But their discourses are structurally and syntactically different.

This difference can, in large part, be attributed to the fact that the environmental justice movement, built as it is on the model of the civil rights movement, has (re)framed the discourse of traditional or reform environmentalism, using what Taylor (2000:515), citing Snow and Benford (1992), has characterized as "elaborated codes . . . which are . . . more inclusive: they are more accessible to aggrieved groups that can use them to express their complaints." The framing of environmental justice has thus created a very accessible *communitarian* discourse that those in disproportionately affected groups can identify with, mobilize around, and, more important, act upon. Within much of the sustainability movement, the discourse is much more academic, requiring *individual* knowledge and skills, and it is therefore less tangible for many who lack the required education and skills.

Lichterman (1995), in his study of the problems inherent in building multicultural alliances between U.S. Green movement groups and antitoxics/environmental justice movement groups, builds on this distinction.

He carried out a two-year participant-observation study on the West Coast of the United States and at regional and national Green and antitoxics conferences. He examined a range of Green groups, including the Seaview and Ridge Greens, "the majority of [whom] were white and had completed at least four years of college," and Hillviewers against Toxics (HAT), "a largely African American group based in a small, predominantly minority populated city" (519).

Lichterman's argument is that Green movement groups are a "personalized form of movement community [that] creates an interdependence of empowered individuals" (515). By contrast, he argues that the antitoxics/environmental justice groups are a "communitarian form of movement community [that] creates an interdependence that weights the group as a whole more, and individuals less" (516). For instance, he notes that "HAT publicly enunciated a collective identity as 'low to moderate income people' in a secular organization fighting corporate incursions. It *carried* that identity as *dutiful members* more than as personally empowered *individuals*" (521; emphasis in original). Similarly, in my interviews with ACE staff and board members, especially those with a history of community organizing and social-justice activism within the African American community, I found that the use of the collective "we" was far more common than the individual "I." Lichterman continues by arguing that in the Green movement literature that he read and in the everyday interactions he observed there was a deep concern for cultural diversity and a sympathy with the racially sensitive environmental justice frame that was being used by antitoxics activists. The caveat here is that while "Green" is a political movement, and antitoxics groups are not *necessarily* synonymous with environmental justice groups, the parallels are close enough for use in my case.

So if Greens, and by tentative extrapolation environmentalists and other adherents to the NEP, are interested in cultural diversity, equity, and justice issues, and African Americans, as one constituent of the environmental justice movement, are as interested as whites in global environmental issues and more interested than whites in neighborhood pollution issues (Mohai 2003), why the gap between the NEP and the EJP?

Lichterman's (1995:527) conclusion is that

> a personalized movement community would result as systematically more accessible to highly educated middle-class groups despite activists' attempts to reach out to diverse communities. . . . this increases the

> probability that the participants will be white. . . . the personalized form of community may be an unrealistic as well as socially limiting basis for a multicultural alliance.

His compromise recommendation for building multicultural alliances, based on Mouffe (1993), is that "some form of shared individualism may be necessary for political community in a society whose members do not all share the same civic traditions and hold different notions of what constitutes 'participation'" (Lichterman 1995:529).

While not offering up the JSP as a panacea for the pain, mistrust, and other barriers to links across the NEP-EJP divide, the crux of my argument is that we simply have to bridge the gap with frank and open discussion, if we are to move toward just and sustainable communities *together.* This can only be done through the concept of just sustainability, whose flexible and contingent discursive frame with its overlapping discourses coming from recognition of the validity of a variety of issues, problems, and framings in many ways reflects Lichterman's compromise of "shared individualism." It is the only common ground. Just sustainability, in this sense, is precisely about movement fusion, which, as Cole and Foster (2001:165) argue,

> is a necessary ingredient for the long-term success of the environmental justice movement because, put simply, environmental justice advocates do not have a large enough power base to win the larger struggle for justice on their own.

I interpret "power base" to mean the totality of resources available to the movement, and this is precisely where the JSP and its organizations —with their discourse of justice (although perhaps not the direct experience of racial, socio-economic, and environmental injustices), their proactive and progressive policies, their "new economics," and their local-global thinking—can work with the EJ movement to build a progressive, multicultural (and intercultural) movement for democratic change. It is to this possibility that I now turn.

Just Sustainability: Global and Local

It is widely known that at the preparatory committees, or "prepcoms," prior to both the UNCED in 1992 and the WSSD in 2002 (and indeed any global environmental or sustainability conference), the usual and ongoing agenda battle happens. This battle is over interpretation and is broadly, but not exclusively, between the richer countries of the North, who want to discuss a "green" agenda of environmental protection, biodiversity, and the protection of the ozone layer, versus the poorer countries of the South, who are proponents of a "brown" agenda of poverty alleviation, infrastructural development, health, and education. McGranahan and Satterthwaite (2000) call these agendas the "ecological sustainability" and "environmental health" agendas, respectively.

This North-South agenda divide is exactly paralleled in the United States by the narrow-focus–broad-focus civic environmentalism divide discussed in chapter 2; that is, the divide between the NEP and the EJP/JSP. Sustainable San Francisco reflects the brown/broad thrust of environmental justice on its website. Using McGranahan and Satterthwaite's (2000) language of environmental health, the website says that "although most environmental justice activists do not use the term 'sustainability' to describe their efforts, for many the survival and environmental health of communities has been a central theme."[2] Guha and Martinez-Alier (1997:21), academics from the South and North, respectively, have argued "'No Humanity without Nature!' the epitaph of the Northern environmentalist, is here answered by the equally compelling slogan 'No Nature without Social Justice!'" (Kothari and Parajuli 1993). This slogan is as compelling in San Francisco, Seattle, or Cleveland as it is in Mumbai, Lagos, or Jakarta.

Another related issue that divides the North and South is the issue of the vast historical "ecological debt" that campaigners from the South argue is owed by the North (Acción Ecológica 1999). Their interpretation is that this debt represents centuries of the appropriation of ecological resources, or "historical" environmental space, such as timber, minerals, and oil from the South, to fuel the industries and line the pockets of wealthy Northerners. As McLaren (2003:30) notes, "even a preliminary assessment of the 'carbon debt' from the imbalance of carbon emissions between rich and poor countries since 1850 suggests that it heavily outweighs the current financial third-world debt."[3]

In short, characterizing the JSP involves taking a broader global interpretive frame than the NEP (and narrow-focus civic environmentalism), on which the traditional and globally dominant Northern agendas are predicated. It involves understanding and supporting both Northern *environment-based* and Southern *equity-based* agendas. As Jacobs (1999:33) argues,

> in Southern debate about sustainable development the notion of equity remains central, particularly in the demand not just that national but that global resources should be distributed in favor of poor countries and people. . . . In the North, by stark contrast, equity is much the least emphasized of the core ideas, and is often ignored altogether.

The green-brown or narrow-focus–broad-focus agenda divide can be demonstrated practically through the issue of urban public transit. Most Northern cities emphasize the environmentally friendly nature of their urban public transit schemes—their ability to get car drivers off the road and their ability to cut pollution loads—with equity issues a lower priority. For example, in Germany, Mayor Beate Weber's Heidelberg City Development Plan 2010 has the following "key objectives of the new Transport Development Plan": reduce environmental burdens, create and preserve liberties, grant the same mobility opportunities to everyone and account for the special situation of persons with mobility handicaps, and reduce dangers and impairments. Equity issues are there, but unlike environmental issues, they are not listed first. (Heidelberg City Development Plan 1999)

By contrast, most Southern city mayors, who are developing innovative schemes such as bus rapid transit (BRT)—such as Enrique Peñalosa of Bogotá or former mayor Jaime Lerner of Curitiba, Brazil—emphasize the *equity* of such schemes, in that car ownership and use is generally the preserve of the rich and BRT schemes allow access to facilities and services irrespective of car ownership.

Again, this is not only a North-South issue but one that is the focus of environmental justice activism in the United States, where transit authorities in many cities such as Los Angeles (see Los Angeles Bus Rider's Union in chapter 4) and Boston (see ACE's Transit Rider's Union in chapter 5) are putting disproportionate resources into affluent suburban areas and commuter services, to the detriment of services in poor inner urban areas. Of course, both environment and equity are important in transit planning and policymaking. I merely use this example to highlight that the sustainability agenda control issue between the North and South is actually an issue between rich and poor everywhere in the world, wherever each may reside.

This kind of analysis, involving ecological debt, environmental space, and other redistributive tools, is fundamental to the JSP for two reasons. First, it focuses the mind on deep structural inequities in the globalized free-trade system (Faber and McCarthy 2003), not just on the usual environmentalist writings about how we need to be more economically and technologically efficient in order to deliver more sustainable communities (Von Weizacker et al. 1996), important though these are. Second, as McLaren (2003:34) argues,

> environmental space (sustainability with equity) and ecological debt (environmental space with history and justice) offer valuable tools, not only to the campaigner and activist, but also to the academic and policy-maker. Fundamentally, they offer a joined-up framework for understanding and promoting both sustainable development and environmental justice.

This framework, and what EcoEquity, a new U.S. organization aimed at advancing the principle of equal rights to global common resources, calls "per capita environmental rights," offer a stark contrast to the "grandfathering" approach to policymaking of the clean-development mechanism and to the joint implementation related to the Kyoto Protocol.

Fundamentally, at global, national, regional, and local scales, the JSP means

> acknowledging the interdependency of social justice, economic well-being and environmental stewardship. The social dimension is critical since the unjust society is unlikely to be sustainable in environmental or economic terms in the long run. (Haughton 1999:64)

Despite this (lack of?) understanding, justice and equity aspects of sustainability are only now beginning to be better understood (Boardman et al. 1999; Agyeman 2000; ESRC 2001; Stockholm Environment Institute 2002; Earth Council 2000; Heinrich Boll Foundation 2002a, 2002b; Agyeman et al. 2003) in terms of environmental space (McLaren et al. 1998; McLaren 2003; Carley and Spapens 1997), ecological debt (Acción Ecológica 1999; McLaren 2003), footprinting (Wackernagel and Rees 1996), and asset-based approaches (Boyce and Pastor 2001).

The discourse of just sustainability is being used to influence policy at the global level and to link global to local. The Earth Charter[4] (2000) represents an initiative to form a global partnership that hopes to recognize the common destiny of all cultures and life forms on earth and to foster a sense of universal responsibility for the present and future well-being of the living world. The Earth Charter Initiative was launched in 1994 by the Earth Council and Green Cross International and is now overseen by the Earth Charter Commission in Costa Rica. The Charter stresses the need for a shared vision of basic values to provide an ethical foundation for the emerging world community (Earth Council 2000). The set of principles that are outlined in the document reflect the necessary and inherent linkages between the ideas of sustainability and justice that will enable the development of this shared vision. The four principles that constitute the basis of the document are respect and care for the community of life; ecological integrity; social and economic justice; and democracy, nonviolence, and peace.

The Stockholm Environment Institute (2002:16) has, through its Global Scenario Group's Great Transition project, begun to map four possible scenarios for the future of the planet: Conventional Worlds, Barbarism, Great Transitions, and Muddling Through. Their preferred scenario, Great Transitions, has two variants: Eco-Communalism and the preferred variant, the New Sustainability Paradigm, which "validates global solidarity, cultural cross-fertilization and economic connectedness while seeking a liberatory, humanistic and ecological transition." This Great Transition, through the New Sustainability Paradigm, is the only one of the four scenarios that sees an increase in equity as essential (Gallopin et al. 1997). In this, the New Sustainability Paradigm is, in Jacobs's (1999) words, "the egalitarian conception"; it moves very close to the JSP.

Similarly, the Heinrich Boll Foundation's "Sustainability and Justice: A Political North-South Dialogue" (2002a) and "The Jo'burg Memo: Fairness in a Fragile World" (2002b), together with some of the Worldwatch Institute's work (strap-line: "independent research for an environmentally sustainable and socially just society") and the World Council of Churches' report "Justice—the Heart of Sustainability" (Robra 2002), explore the more global aspects of just sustainability. All of these documents offer a global-to-local context to U.S. efforts toward just sustainability.

Some authors, however, such as Dobson (1999, 2003), take a separatist and traditional environmentalist view. They argue that the concepts of and movements for sustainability and environmental justice will come into conflict because of the environmental justice movement's primary focus on the issue of social equity, whereas the focus of environmental sustainability[5] is on green issues. As I have argued elsewhere (Agyeman et al. 2003) and in earlier chapters, the wider discourse and activism of broad-focus civic environmentalism and the JSP encompasses a far more expansive set of policy goals and social groups than that of environmental sustainability, as evidenced through the NEP (see table 3.1), whereas Dobson (1999, 2003) does not appear to see sustainability in anything other than environmental terms.

I term the proponents of strictly environmental sustainability "unreconstructed environmentalists" (Agyeman 2001). In Harvey's (1996: 391–392) well-chosen words, they are "that wing of environmentalism that focuses primarily on 'nature' as wilderness, species, and habitat preservation." Interestingly, having first identified their cause as "environmentalism," Harvey then goes on to identify the protagonists in this wing not as environmentalists but as "ecologists." He continues that "theoretically, ecologists claim that everything is related to everything else, but they then marginalize or ignore a large segment of the practical ecosystem" (392). The "large segment" of the practical ecosystem is interpreted here as being humans, and it is easy to see why, in reaction to the nature-focus of traditional environmentalism, the EJ movement has been accused of being anthropocentric.

I accept that the environmental justice discourse and movement has addressed a more clearly defined, specific constituency marginalized by both race and income. However, this does not, as Dobson (1999, 2003) suggests, preclude common focuses and interests, as I demonstrate later in this chapter and in chapter 5. Indeed, Schlosberg (1999:194), in his investigation of the prospects for a critical pluralism, argues that there is a growing number of

> examples of cooperative endeavors between environmental justice groups and the major organizations. The key to these relations is an understanding of the justice of environmental justice on the part of the major groups, and an attention not just to the end goal of a particular environmental agreement or policy, but to the process of such a battle.

Three points are worth making here. First, it is precisely the "justice of environmental justice" that the JSP has adopted from the EJP, which shows its ability to develop overlapping discourses. This flexibility

TABLE 3.1
The Just Sustainability and New Environmental Paradigms Compared

	Just Sustainability Paradigm	New Environmental Paradigm
Main contributors	Agyeman et al. (2003); Haughton (1999); Campbell (1996); Shutkin (2000); Warner (2002); Roseland (1998); Middleton and O'Keefe (2001); ESRC (2001); Stockholm Environment Institute (2002); McLaren (2003); Dunion and Scandrett (2003); Schlosberg (1999).	Catton and Dunlap (1978); Dunlap and Catton (1979); Milbrath (1989).
Central premise and focus	The interdependence of social justice, economic well-being, and environmental stewardship is a prerequisite in developing sustainable communities.	Better environmental stewardship from ecocentric or technocentric[a] position.
	Focus on quality of life; present and future generations; justice and equity; living within ecosystem limits.	Focus on quality of life; present and future generations; living within ecosystem limits.
Approach to civic renewal and engagement	Aim is to create an inclusive, representative, deliberative civic process using deliberative and inclusionary processes and procedures (DIPS) such as citizen's juries, Future Search, and visioning conferences that serve to actively engage the public in all areas of policy formation and implementation. In other words, *process* is as important as *product.*	Because the premise is environmental stewardship, the NEP's approach is limited to environmental aspects of renewal and engagement. Participation is limited, based on rounding up the usual suspects, as *product* is more important than *process.*
Policy-based solutions	As a politically transformative paradigm, the JSP looks toward a participatory democracy. However, in the transition period, it looks to joined-up or connected policy: integrated social-economic-environmental policymaking based on principles such as demand management and resource decoupling.	As a reformist paradigm, it has a focus on policy change (technocentric wing) as a solution to problems. Ecocentrists are more like JSP on policy-based solutions.
Attitude to planning	APA's (2000) *Planning for Sustainability,* about planner's role in sustainable futures, was a good start, but little focus on justice and equity issues. One of the four main objectives of APA policy, "fair and efficient use of resources" is not strong enough.	APA (2000) guidelines are strong on NEP concerns. For instance, three of four objectives of APA policy are environmental, such as fossil fuels, chemical accumulation, reconcentrating and reconstructing wastes.

(continued)

TABLE 3.1 *(continued)*

	Just Sustainability Paradigm	New Environmental Paradigm
	Comprehensive plans should include just sustainability considerations.	
Policy tools[b]	Environmental space, ecological debt, eco-tax, elimination of agricultural and energy subsidies, affordable housing, LETS, recycling and renewable energy, efficient and integrated transportation systems, Community Supported Agriculture, sustainability indicators, environmental justice indicators (Harner et al. 2002), sustainability inventory, sustainability appraisal, precaution, ISEW/GPI. Like EJ, adopts community approaches to risk assessment and research.	Use of technico-scientific tools, such as EIA, often expert led. Eco-tax, elimination of agricultural and energy subsidies, LETS, recycling and renewable energy, efficient and integrated transportation systems, Community Supported Agriculture.
Attitude to markets and economy	Call for a "new economics" that says after a certain amount, more money (standard of living) does not mean more happiness (quality of life). Markets, where they are the correct mechanism, should be imbued with values that ensure they work for the common good. Sufficiency is as important as efficiency. Strong on worker rights, corporate accountability.	"New economics" is the goal for ecocentrists, but technocentrists favor the neoliberal status quo; efficiency over sufficiency.
Difference from Environmental Justice Paradigm (Taylor 2000)	Similar in many senses in that the focus is on justice and equity. Differences are that (1) the JSP has a central premise on developing sustainable communities; (2) despite both being transformative paradigms, the JSP has a wider range of progressive, proactive, policy-based solutions and policy tools; (3) the JSP is calling for and has developed a coherent "new economics"; (4) the JSP has much more of a local-global linkage; (5) the JSP is more proactive and visionary than the typically reactive EJP.	Many differences based on the fact that justice and equity are only just implicit, not explicit. For a full comparison see Taylor (2000:543–545).

[a] Ecocentrism and technocentrism (O'Riordan 1970) are orientations of environmentalism reflecting nature-focused, or technology-focused, pathways or futures. Ecocentrism relates to strong sustainability, while technocentrism relates to weak sustainability.

[b] Only a few of the JSP policy tools are mentioned here.

comes from recognition of the validity of a variety of issues, problems, and framings. This makes the JSP a very different paradigm from the NEP or from the environmental-stewardship-oriented sustainability of "the major groups" that Dobson (1999, 2003) discusses.

Second, "justice" is also the bridge over the "equity gap" between the EJP and NEP. It could also be seen as a "welcome area" in the free spaces I mentioned at the beginning of this chapter. Justice and equity will therefore be a critical focus in developing both more cooperative endeavors, and, most importantly, movement fusion.

Third, the focus on *processes* as well as *products* is well made. Many groups, projects, and initiatives within the NEP, such as CBEP projects, believe that they are doing the right thing, working for a *substantive* solution—a *product*. And of course they are. That is what most of their mandates require them to do. However, if they want a more diverse base of support, as most say they do, they must look to the *procedures* or *processes* by which they achieve their mandated aims. Groups operating within the JSP are deeply concerned about the *destination*, a just and sustainable community or society, but they are equally concerned about the *journey*, as evidenced through their pioneering use of DIPS.

Schlosberg (1999:194) builds on his concept of cooperative endeavors to offer us a just environmentalism. He says that such cooperative endeavors "offer the environmental movement as a whole a way out of its limited and conventional pluralist approach, into a realm of more diverse, participatory, effective and just environmentalism." In this, he is clearly critiquing the discourse of narrow-focus civic environmentalism and the NEP and is positing a notion akin to just sustainability, although he does not develop it. Longo (1998), while sympathetic to cooperative endeavors, takes a more utilitarian view. He argues that environmental justice groups can provide traditional environmental groups "with a powerful and immediate electoral advantage at the local level," that they can "enhance the national reputation, and thus legitimacy of mainstream groups," and that they can "add to the membership rolls of mainstream groups" (173). This may be so in some cases, although my work with ACE (chapter 5) shows that coalition building is not as easy or as straightforward as one might think, even given a common focus; many other issues come into play.

Local cooperative endeavors (Schlosberg 1999) between environmental justice and sustainability groups are the small building blocks that are needed to build up to Cole and Foster's (2001:164) movement fusion: "the coming together of two (or more) social movements in a way that expands the base of support for both movements by developing a common agenda." This cooperation, co-activism, and co-strategizing on sustainability and environmental justice issues, for it cannot truly be called movement fusion yet, can be found in local fights for, and conferences on, just transportation (for instance, Conservation Law Foundation 1998; Bullard and Johnson 1997; Bullard et al. 2004), community food security (Gottlieb and Fisher 1996; Perfecto 1995), smart growth, sustainable communities and cities (Beatley and Manning 1997; Roseland 1998; Rees 1995; Haughton 1999; Bullard et al. 2000), the Precautionary Principle, and clean production (*Rachel's Environment and Health News* 1998, 1999), among others.

The Just Sustainability Paradigm

The discourse of the JSP, and the Just Sustainability Index that I introduce in the next chapter, flow from the definition of sustainability of Agyeman et al. (2003:5): "the need to ensure a better quality of life for all, now and into the future, in a just and equitable manner, whilst living within the limits of supporting ecosystems."

In this definition are four, albeit related, focal areas of concern:

- Quality of Life
- Present and Future Generations
- Justice and Equity
- Living within Ecosystem Limits

Quality of Life

As was argued in the previous chapter, GDP, the headline indicator of standard of living, does not account for human activities like domestic and family tasks, taking care of kids or the elderly, volunteer or community work, and leisure-time activities. GDP is therefore only a measure of monetary wealth, that is, standard of living, not of unpaid human activities that develop social capital and help improve our quality of life. In the previous chapter, I gave the ISEW as an example of an alternative index of progress, another way of looking at development. More

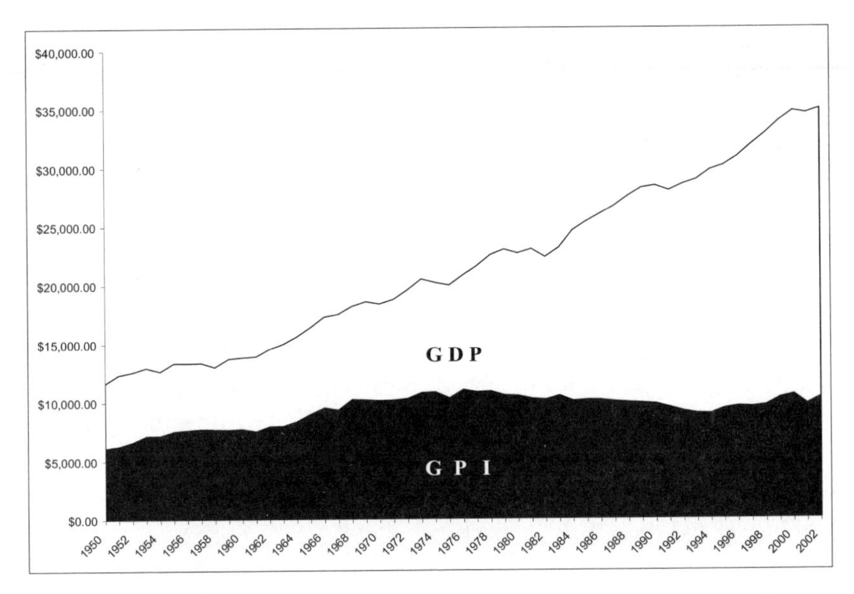

Fig. 3.1. Gross Domestic Product versus Genuine Progress Indicator for the United States, 1950–2002, Per Capita (from Redefining Progress, "The Genuine Progress Indicator, 1950–2002 [2004 Update]," March 2004).

commonly used in the United States is the Genuine Progress Indicator (GPI), which is constructed in a similar way to the ISEW.

Redefining Progress (2004) notes,

> the GPI uses more than 20 aspects of our lives to evaluate the economy. It begins with the same personal consumption data that GDP is based on, but makes additions for positive contributions (such as volunteer work and housework) and deductions for negative activities (such as crime, underemployment, and the degradation of natural resources). The result is an economic indicator that gets much closer to the economy that people experience.

It is the gap between our GDP or quantitative wealth, which, as fig. 3.1 shows, is rising for many people, and our GPI or qualitative wealth, which is falling, that explains the lack of a "feel good factor": we have more money in our pockets, but we are less happy. In the words of Cobb, Halstead, and Rowe, in the October 1995 edition of the *Atlantic Monthly,* "If the GDP Is Up, Why Is America Down?"

Present and Future Generations

Haughton (1999:65) has developed an approach that lists five equity and justice principles that he argues "make a concern for sustainable development different from the existing concerns of environmental planning." These principles are intergenerational equity (the futurity principle), intragenerational equity (social equity or social justice), geographical equity (or trans-boundary responsibility), procedural equity (open and fair treatment), and interspecies equity. In effect, Haughton is signaling the difference between the JSP and traditional environmentalism. The NEP has a position on each of these equity and justice principles except intragenerational equity, which is poorly developed or nonexistent within the NEP. Yet intragenerational equity is the reason for the development of the environmental justice movement and is of pivotal importance in the development of sustainable communities.

Justice and Equity

In the same way that the environment is pivotal to the NEP and to environmental sustainability, justice and equity are pivotal in the JSP. Justice is, in the JSP, not simply about procedural justice or access to decision making and decision makers, it is also about distributive and substantive justice: affordable housing, a cleaner environment, and so on.

Indeed, Roberts (2003:230) argues that

> the reduction of social exclusion and the promotion of greater social cohesion and justice are fundamental objectives that are as important to achieving sustainable development as responsible economic progress and the effective management of the environment.

Low and Gleeson (1998:14) argue that "sustainable development without environmental justice is an empty formula." However, they also talk of *ecological justice,* which they argue is "justice to non-human nature" (157), as a key partner to environmental justice, which they see as "the distribution of environmental quality" (133).

Their point is that we must find ways to expand the dialectic of justice to ecological justice and that this requires a shift in our thinking about the nonhuman world from instrumental to moral and a shift in the "way we think of our 'self' and thus how we define our interests and moral values" (133). It is easy to see, though, why, despite the inclusion of nonhuman nature in the Principles of Environmental Justice (see Appendix), the movement has been accused of being too anthropocentric. Its anthropocentrism is a reaction against decades of environmental activism focused on nature while neighborhoods of color and low income, like those in Roxbury and Dorchester in Boston where ACE and DSNI are active, were disinvested, redlined, and saturation patrolled by the police.

The JSP is increasingly questioning the justice and equity implications of international agreements, especially those related to trade or economic development. There is great (and under-researched) potential for the notions of environmental justice, human rights, and sustainability to permeate environmental regimes and international policy and agreements. Indeed, it is being increasingly recognized that one of the best ways to protect environmental rights is to uphold the basic civil and political rights of the individual (Sachs 1995). Furthermore, once these ideas become enshrined in policy, they have the capability to enable legal challenges to existing practices on which they could not make a previous impact. An example of this confluence of activism, policy, and law can be seen in recent developments surrounding the use of Title VI of the 1964 Civil Rights Act to prove discriminatory intent in environmental injustice cases. However, the difficulty of conclusively proving "racial intent" has hampered activists who have focused on such violations. Cutter (1995) notes that because of this difficulty in using the equal protection doctrines, toxic torts are a much more likely remedy, as civil rights laws are not as effective as environmental or public health laws.

Living within Ecosystem Limits

Tools like the ecological footprint (Wackernagel and Rees 1996), but more so ecological debt (Acción Ecológica 1999) and environmental space (Spangenberg 1995), with their basis in the equitable use of resources, are providing us with a more accurate map as to what *justice*, in the context of living within ecosystem limits, really means. In sustainability terms, we have never been here before, nor have we had such metrics. Our current Northern consumption rates of essential resources need to be reduced in the order of 80 percent to achieve sustainable consumption. For example, McLaren et al.'s (1998) environmental space targets for the United Kingdom (there are, as yet, no U.S. calculations, for reasons I put forward in chapter 2) vary from a reduction of 15 percent for water, 88 percent for aluminum and carbon dioxide, to 100 percent for chlorine. This is far greater than the less than 10 percent cuts that are being discussed in many future-resource scenarios, including carbon dioxide production.

Serious questions also arise about how we move the public toward an awareness of the need for these reductions, to engender support for them, and to take action toward them. At the local, state, and federal levels, progressive policies must be developed, like those discussed in chapter 2, which can both manage demand for a resource, rather than increase its supply, and decouple increased resource consumption from well-being.

With regard to managing demand, the 2003 North East American blackouts, which affected 25 percent of the continent's population, starkly exposed the difference between U.S. and Canadian attitudes toward energy production and consumption. While the U.S. radio stations were talking about the need to increase supply so that such an occurrence never happens again, the Ontario Provincial Government in Canada was asking its citizens to manage their demand by using electricity sparingly and at off-peak times.

With regard to decoupling increased resource consumption from well-being, the current and increasingly linked policy debates around growing U.S. car usage, sprawl, obesity, and general health, and the need for Americans to eat far less for a healthy life, are examples of how increasing resource consumption can have a negative effect on well-being and quality of life (Wilkinson 1996). In the words of Stutz and Mintzer (2003:51), "having shown that consumption will not increase well-being after our basic needs are met, we must focus our energy on cultivating other aspects of a good life in order to profoundly improve our quality of life." The "other" aspects they cite are "community, equality, and leisure" (51).

Both the demand-management and resource-decoupling policy debates are beginning to raise the profile of a concept at the core of the JSP discourse: sufficiency. This is the equity-based sustainability message that less can be more. It will be of increasing importance in the coming years, especially in the North, as we begin to develop demand-management policies in order to limit our resource consumption by those amounts suggested by McLaren et al.'s (1998) environmental space calculations, so that Southern countries can consume their fair share of environmental space—commensurate with improving their standard of living—and, thereby, their quality of life. *Sufficiency* complements but also contrasts with the environmentalist-based sustainability concept that runs through the heart of the European environmental modernization agenda: *efficiency*, or doing more with less.

Progressive policy ideas based around these concepts, such as those in the Proposed Principles of Environmental Justice of Boardman et al. (1999),[6] were developed in the United Kingdom, but as policy principles they are relevant to U.S. activists and academics. The principles of Boardman et al. are more policy ready than the 1991 Principles of Environmental Justice and are, in effect, a coherent and thought-provoking agenda for just sustainability in the United Kingdom. The overall message of the principles is loud and clear. If we are serious about just sustainability,

> revenue needs to be raised or redirected for capital investment, pricing structures need to give incentives and be progressive, the market needs to be transformed to deliver change more quickly and any remaining inequality problems offset through compensation. (Boardman et al. 1999:24)

Developing Public Support

If the wider public is to support such progressive policies, there must be an entitlement to a cross-curricular "education for sustainability" in schools and universities and a broadly based, accessible popular education campaign. However, purely information-based campaigns, whether by government or by not-for-profits, generally fail to engender action, unlike those based on social marketing techniques[7] and those using DIPS (Kollmuss and Agyeman 2002).

Social marketing is derived from commercial marketing and behavioral psychology and can be used to encourage new (healthier, more sustainable, environmentally friendly) behaviors among groups of people. Social marketing techniques have been widely used in the field of public health, antismoking campaigns, AIDS awareness campaigns, and to encourage the treatment of leprosy. The development of Community-

Based Social Marketing (CBSM) for sustainability arose out of concerns about the ineffectiveness of campaigns that relied solely on providing information about the negative impacts of unsustainable behavior. These so-called information deficit campaigns (Burgess et al. 1998: 1447) have failed to create the kind of transformative behavior that is required to move societies toward environmental sustainability, never mind the qualitatively and politically different proposition of just sustainability.

The pragmatic approach of social marketing has been offered as an alternative to information deficit campaigns and has been shown to be very effective at bringing about behavior change (McKenzie-Mohr and Smith 1999). CBSM identifies the benefits and barriers to behavior and then organizes the public into groups that have common characteristics, so that the delivery of programs can be made most efficient. McKenzie-Mohr and Smith (1999) claim that the primary advantage of social marketing over other forms of community education is that it starts with people's behavior and works backward to select a particular tactic suited for that behavior.

Although the concept of CBSM is relatively new, there exists a set of tools for implementing a social marketing program that could complement progressive policymaking. Some of the social marketing tools include vivid personal communication, prompts, obtaining a commitment, and norm appeals. These have been used by Global Action Plan (GAP) as part of its domestic Eco-Team program, designed to help "bring your household into environmental balance" (Gershon and Gilman 1990:2). One criticism of GAP's work has been that the Eco-Team program works well in middle-class, high-ownership areas but because of lower home-ownership rates is not appropriate for low-income neighborhoods. To counter this, the Empowerment Institute, in collaboration with the City of Philadelphia, developed the Livable Neighborhood program, which focuses more on the neighborhood than on the home.

I am not suggesting that CBSM is a panacea, a harbinger of just sustainability. That requires a paradigm shift, a transformation of the status quo. CBSM is, however, a form of community-based learning that seems to deliver results in developing pro-environmental behaviors (Kollmuss and Agyeman 2002), for example, increasing public transit use (Bachman and Katsev 1984) and decreasing energy usage in student dormitories (Marcell et al. 2004).

Just Sustainability and Environmental Sustainability

The JSP, into which I would place broad-focus civic environmentalism, together with the more socio-politically progressive elements of the NEP/sustainability movement, combines and extends some key features of both the EJP and NEP. At the same time, it develops some unique new ground that makes it and the organizations within it more effective in the pursuit of just and sustainable communities than the EJP alone. This new ground is detailed in the five numbered sections that follow.

Crucially, however, the JSP does not supplant the EJP but is operative alongside it, with their discourses overlapping. They are complementary. The JSP represents, in many ways, a bridge between the EJP and NEP. As such, the JSP is an acknowledgment of the successes of the EJ movement in getting justice on the environmental agenda ("the justice of environmental justice," Schlosberg 1999) and of the failures of the NEP to develop a realistic, justice-based political project. At this stage it is worth making clear that the interpretive differences (i.e., in core values and beliefs, environmental philosophy, political ideology, diagnostic attribution, and repertoire of action) between the JSP and NEP (especially the technocentric wing) are greater than those between the JSP and the EJP. In summary, the foundation of the JSP and EJP in justice and equity discourses has led to both having a more politically aware analysis than the NEP. Looking in more detail at the five major differences between the JSP and the EJP will assist in clarifying the emergence of the JSP and how it complements the EJP rather than detracts from it.

1. The JSP Has a Central Premise on Developing Sustainable Communities

There is a vast, detailed, and growing theoretical and practical literature from North America (Medoff and Sklar 1994; Kline 1995; Campbell 1996; Beatley and Manning 1997; PCSD 1996, 1997; Roseland 1997, 1998; Hempel 1999; Mazmanian and Kraft 1999; Shutkin 2000; Agyeman and Angus 2003; Portney 2003), from Europe (Lafferty 2001; Barton 2000), and more globally focused (Satterthwaite 1999; Newman and Kenworthy 1999; Pugh 2000) on the development of sustainable communities and cities. This literature has detailed the visions, tools, techniques, and strategies for getting there, described the use of indicators for telling us if we are there, and given cases of communities moving toward sustainability.

There was a flurry of EJ movement activity on the vision of sustainable communities, especially in the 1990s. Events such as the 1994 conference funded by the Tides Foundation, "Defining Sustainable Communities: Many Pieces Fit Together"; the 1994 conference organized by the South West Network for Environmental and Economic Justice (SNEEJ), "Dialogue: Building Sustainable Communities"; the 1995 "Sustainability and Justice Conference and Reader," hosted by the Urban Habitat Program in San Francisco; EPA's (1996) report "Environmental Justice, Urban Revitalization and Brownfields . . . Envisioning Healthy and Sustainable Communities"; and finally, the 1997 Environmental Justice Resource Center (EJRC) conference, "Healthy and Sustainable Communities: Building Model Partnerships for the 21st Century." The EJRC conference built on the successes of those conferences previously mentioned in order to articulate some general principles of sustainable development from the perspective of the EJP:

- Grassroots community groups want to see sustainable development that is not only economically and ecologically sound but is also just.[8]
- They support a sustainable economy that improves the vitality and self-sufficiency of their community and its residents.
- They view education as a key ingredient in long-term community health and sustainability plans.
- They advocate the right of all people to a safe and secure livelihood, including the right to education, safe and affordable housing, and adequate health care.
- They promote democratic access to and control over natural resources.
- They demand that all groups are included as equal partners in development decisions.
- They promote government and corporate accountability to the public for decisions about production and consumption.
- They support the acquisition and preservation of open space in our community.
- They promote respect for cultural diversity, Mother Earth, and the spiritual connectedness among all living beings.
- They support the right of all workers to a safe and healthy work

environment, without being forced to choose between an unsafe livelihood and unemployment.
- They support public policy decision making based on mutual respect and justice for all people, free from any form of discrimination. (EJRC 1997:4)

Two promising developments, not from within the EJ movement but with obvious sympathies, are the DSNI revitalization efforts and the creation, in March 2001, of the Maryland Commission on Environmental Justice and Sustainable Communities (Executive Order 01.01.2001.01). Effective October 1, 2003, the commission was established by statute (Chapter 460, Acts of 2003). The commission advises state agencies on issues related to environmental justice and sustainable communities, it analyzes and reviews what impact current state laws, regulations, and policy have on the equitable treatment and protection of communities threatened by development or environmental pollution, and it determines what areas in the state need immediate attention. With the Children's Environmental Health and Protection Advisory Council, the commission coordinates recommendations on such issues. The commission will assess the adequacy of current statutes to ensure environmental justice and, like Massachusetts's Environmental Justice Populations, develop criteria to pinpoint which communities need sustaining.

Despite a productive phase in the 1990s, however, the EJ movement has not, other than the principles listed above, articulated a coherent, replicable policy agenda for sustainable communities, which like the Principles of Environmental Justice is pivotal to its work. Similarly, it has not *collectively* endorsed any of the versions currently available, from the efforts of the PCSD in its 1997 *Sustainable Communities* report (although EJ leaders were involved) to those available at present. Interestingly, a lot of rationales from the EJ movement in the 1990s for sustainable communities were economically based. Gross (1997:19) is typical: "at the heart of sustainable communities are economies that are able to sustain those communities."

As I show in chapter 5, the reason there was a considerable amount of EJ activity around sustainable communities in the mid-1990s was that it was purely a response—a reaction to the PCSD setting the agenda. This is not a criticism but a political reality in a movement the grassroots of which has been overwhelmingly reactive in the face of massive adversity. It is also clearly an opportunity for the EJ movement and should represent, as Goldman (1993) has suggested, a part of the next phase of its development.

2. The JSP Has a Wider Range of Progressive, Proactive, Policy-Based Solutions and Policy Tools

These progressive, equity, or fair-shares tools such as environmental space, ecological debt, and eco-tax reform, together with deliberative tools (DIPS) such as Future Search, consensus conferencing, and study circles (table 3.1), are beginning to represent a coherent policy architecture, a viable alternative to grandfathering and other inequitable approaches to policy, characteristic of the neoliberal status quo, as represented by the Dominant Social Paradigm (DSP) of Milbrath (1989). The development of new tools and policies is an area in which the EJ movement has been less active, except for its advocacy for state-based EJ and other related policies, such as Areas of Critical Environmental Justice Concern (ACEJC) in Massachusetts, and its democratizing approaches to tools such as risk assessment, GIS, and research.

However, Cole and Foster (2001:112) note the recent move away from pluralism as the preeminent model of public participation, toward an increased use of deliberative structures "in many state permitting processes," and they note that

> local advisory committees offer the opportunity for a qualitatively deliberative process, one that creates an opportunity for lay and technical people to work together, have a dialogue, and reach consensus.

The kudos for this shift rests in part with EJ activists.

3. The JSP Is Calling for, and Has Developed, a Coherent "New Economics"

Goldman (1996:129–130) acknowledges the value of the EJ analysis in local racial disproportionality; however, he notes that there is more to be done domestically and internationally with regard to economic forces:

> the most prominent members of the environmental justice movement focus on the racial dimensions of disproportionate environmental im-

> pacts—and with good reason, since these disparities represent the most vile symptoms of injustice. But the leadership of the movement has only begun to scratch the surface of the complex global and domestic forces that underlie the paramount trend of increasing inequality.

This global, and domestic or local, analysis, together with tools to deal with the problems, such as environmental space and ecological debt, is precisely where the JSP is particularly strong. In developing a "new economics" predicated on the notion of *sufficiency*—that is, more money (standard of living) does not mean more happiness (quality of life)—the JSP is not saying that the poorest people and communities, like Massachusetts's Environmental Justice Populations or countries in the South, should languish in material poverty while being ostensibly happy. It makes explicit that the richest people and countries "have the infrastructure, the capital and the educational and social investments necessary to allow them to bear a much greater share of the transitional costs" (McLaren 2003:30). At the national scale, bearing of transitional costs, both by richer communities in the United States and, globally, by richer countries such as the United States, is clearly a political time bomb, but it is one that we will surely have to deal with if humankind is to move toward more just and sustainable communities, societies, and futures.

4. The JSP Has Much More of a Local-Global Linkage

The Earth Charter, as mentioned earlier, is a global charter of just sustainability. It has been adopted by more than thirty cities across the United States including Berkeley, California, and Burlington, Vermont; by the U.S. Conference of Mayors (one thousand members); and by the Florida League of Cities (four hundred members). Furthermore, while the EJ movement has sought to establish links and global dialogues, tools like environmental space and ecological debt provide the JSP with an opportunity for real, robust, and just resource-allocation strategies and policies at global, national, and local levels. As McLaren (2003:34) argues,

> the concentration of toxic contaminated sites in American black communities is an expression of the same unsustainability as the expropriation of indigenous subsistence resources by logging companies in

> Indonesian forests. Both are facilitated by the processes of liberalization and globalization that have shaped economies—national and global—over the past decades.

This local-global reflexivity is precisely why Friends of the Earth Scotland, a country not traditionally known for EJ issues, has constructed a campaign for environmental justice that may be a model for other countries and groups to follow. It uses an adaptation of Carley and Spapens's (1997) notion of "equal distribution of resource consumption between countries on a per capita basis" and is based on the concepts of environmental space and ecological debt. Like the work of Boardman et al. (1999), Friends of the Earth Scotland's (FoES) campaign is called "environmental justice," but it falls squarely into just sustainability. As Dunion and Scandrett (2003:312) note, "in our usage, environmental justice combines challenging the unequal social distribution of poor environments, with the requirements to meet resource reduction targets determined by environmental space." The campaign's launch, with the ethical slogan "no less than our right to a decent environment; no more than our fair share of the Earth's resources," coincided with the creation in 1999 of the Scottish Parliament in Edinburgh. The parliament has the legislative power and capacity to set an agenda through guidance to local authorities, develop voluntary agreements, and provide direction to quangos (quasi-autonomous non-governmental organizations). At the launch, FoES's director at the time, Kevin Dunion, said, "I shall be calling for the new Scottish Parliament to show that it is serious about making real change. We want targets for improving energy efficiency in industry; an energy rating for all homes within 10 years; a Warm Homes Act to eradicate fuel poverty; national and local targets under the Road Traffic Reduction Act; and changes to Scottish building regulations to improve energy performance" (FoES 1999). These targets, among others, now form a part of FoES's Environmental Justice Action Plan (FoES 2000).

Another innovation is FoES's collaboration with Queen Margaret University College in Edinburgh, which provides cutting-edge courses in environmental justice, designed for people active in working for environmental justice in their community or workplace. Courses currently are run in two forms: as a community-based version (through residential teaching) and a workplace-based version (through web-based distance learning).

What distinguishes the Scottish EJ model from those in the United States, which tend to be more local, with regional networks of differing vitality, is its heightened sense of global responsibility and equity on a foundation of sufficiency: "no less than our right to a decent environment; no more than our fair share of the Earth's resources." This is not to say that either U.S. or Scottish models are right or better but that the local-global linkage of the Scottish model offers an international contextualization, which is certainly something that the U.S. EJ organizations, and movement as a whole, could work on.

The problem we have in the United States, and in North America generally, is that many of the progressive just sustainability concepts and tools, like the Precautionary Principle before the 1998 Wingspread Conference in Racine, are not yet being evaluated here, as they are in countries like Denmark, The Netherlands (Sustainable Netherlands), the United Kingdom (Friends of the Earth England, Wales, and Northern Ireland), and Europe as a whole (European Environment Agency, Wuppertal Institute, Sustainable Europe Research Institute), which have environmental-space calculations and are exploring their use as policy guides.

This seeming lack of interest in the United States is precisely because of the points made earlier by Jacobs (1999) and McLaren (2003): these tools change the agenda from *environment* to *environment and equity,* with all the political consequences that such a shift implies. In addition, as I mentioned in chapter 2, a limits-based discourse in the United States is a very difficult sell. This also explains why the ecological footprint concept is popular in the United States. It tells us that we are living unsustainable lifestyles in the North by showing us the land area required to sustain those lifestyles, but unlike environmental space, which is a much better, more powerful policy tool, it does not say specifically how much less we should consume of any given resource.

5. The JSP Is More Proactive and Visionary Than the Typically Reactive EJP

The EJP, or more correctly, the grassroots environmental justice movement, has typically been reactive in the light of external threats such as toxic dumping, TSDFs, and other LULUs. There is no surprise or shame in this. Grassroots environmental justice groups are often lacking in their ability to frame the issue, seize on political opportunities, and mobilize the political and financial resources needed to be more proactive, that is, heading off problems before they arise. Wider adoption of the Precautionary Principle, with its changed conception of the burden of proof, will undoubtedly help in this as, and if, it is more widely adopted in the United States.

The JSP uses deliberative tools to help communities develop common values and visions, agree on action plans with targets, and monitor progress toward community-agreed goals. This proactivity in saying what communities want, rather than being reactive and saying what they do not want, is a much better position around which to articulate policy demands.

Conclusion

In this chapter, I have characterized the JSP. I have shown its differences from the current sustainability paradigm (the NEP) and the EJP. These differences are what I believe makes the JSP a more flexible, fundamentally transformative paradigm than the NEP, but I also believe that the emergent JSP exists alongside the EJP and that they can coexist, as my case study of ACE in chapter 5 shows. In this sense, the JSP is not looking to take over the EJP or EJ movement. Rather, I see the two working synergistically, recognizing the validity of a variety of issues, problems, and framings in a range of local coalitions. In the next chapter, I want to show some examples of communities and organizations that are utilizing some of the concepts, if not the tools and techniques, of the JSP.

4

Just Sustainability in Practice

This chapter moves the theory of just sustainability into practice.[1] First, I develop a Just Sustainability Index, through which I assess the commitment of a range of national environmental and sustainability organizations to the JSP. This is done in order to provide a rough and ready metric, a rule of thumb as to where well-known national organizations stand in relation to justice and equity issues. I then present three representative programs or projects in each of five sustainability issue categories (land-use planning, solid waste, toxic chemical use, residential energy use, and transportation) that are demonstrating just sustainability in practice in U.S. cities.

The Just Sustainability Index

In order to chart the current status of the just sustainability discourse and of the JSP among national environmental and sustainability membership organizations in the United States, a selection of international organizations, and programs and projects in U.S. cities, I developed a Just Sustainability Index (JSI) as a hybrid of discourse analysis, content/relational analysis, and interpretive analysis. The JSI uses the categories listed in table 4.1.

Using organizational websites and the search terms "equity," "justice," and "sustainability," I looked at both organizational mission statements and prominent contemporary textual or programmatic material. Derivations of equity, justice, and sustainability, such as "equitable," "just," and "sustainable," were also used if the original terms yielded no results. In addition, to fully ensure that no organization was potentially excluded, sentiments such as "the fundamental right of all people to have a voice in decisions," "disproportionate environmental burdens," and mention of "environment" instead of "sustainability" (only if

TABLE 4.1
The Just Sustainability Index

0	No mention of equity or justice in core mission statement or in prominent contemporary textual or programmatic material.
1	No mention of equity or justice in core mission statement. Limited mention (once or twice) in prominent contemporary textual or programmatic material.
2	Equity and justice mentioned, but focused on intergenerational equity in core mission statement. Limited mention (once or twice) in prominent contemporary textual or programmatic material.
3	Core mission statement relates to intra- and intergenerational equity and justice and/or justice and equity occur in same sentence in prominent contemporary textual or programmatic material.

associated with "justice" or "just") were counted as having fulfilled the search terms. Reliability was ensured by having an additional researcher code the organizations.

As I mentioned in the book's introduction, the JSI comes with some caveats and limitations. If I only looked at organizations' statement of their mission, I could be accused of not actually getting at behavior, merely textual representations of reality and symbolic declarations. That is why I look at both "mission" and "program" issues, since most organizational websites, certainly those of the organizations I dealt with, have a wealth of up-to-date programmatic information. This programmatic information, in combination with the mission statement, provides a relatively accurate picture of an organization's commitment to the JSP.

The choice of which organizations to survey, it could be argued, is somewhat arbitrary. No official list of national environmental and sustainability organizations exists. Many of the organizations that I surveyed (see table 4.2) were derived from SaveOurEnvironment.org, a collaborative effort of the nation's most influential environmental advocacy organizations including all the Big Ten groups. From these groups, a "snowball" technique was applied in order to find more organizations.

Three conclusions can be drawn from the results of my survey. First, among the thirty national environmental and sustainability membership organizations selected in my survey, more than 30 percent had a JSI of 0. This means that in such organizations there is no mention of equity or justice in their core mission statement or in prominent contemporary textual or programmatic material.

Second, the average JSI was 1.06. While not statistically significant, this suggests that the majority of U.S. national environmental and sustainability membership organizations make no mention of equity or justice in their core mission statements and limited mention (once or twice) in prominent contemporary textual or programmatic material. This backs up Taylor's (2000) point about the lack of interest in social justice within the NEP.

Third, only organizations with a JSI of 3 could be considered to have more than a passing concern for just sustainability and be operating within the JSP. In other words, their core mission statement relates to intra- and intergenerational equity and justice and/or justice and equity

TABLE 4.2
Just Sustainability Indices for National Environmental/Sustainability Organizations Requiring Membership

Organization[a]	Just Sustainability Index
American Rivers	0
American Solar Energy Society	0
Center for Health, Environment and Justice	3
Center for a New American Dream	3
Defenders of Wildlife	0
Earth Island Institute	2
Earthjustice	2
Environmental Defense	3
Environmental Law Institute	1
Friends of the Earth	2
Greenpeace	1
Izaak Walton League	1
League of Conservation Voters	0
National Audubon Society	0
National Environmental Trust	0
National Parks Conservation Association	1
National Wildlife Federation	0
Natural Resources Defense Council	2
Nature Conservancy	0
North American Association for Environmental Education	2
Ocean Conservancy	0
Physicians for Social Responsibility/EnviroHealthAction	1
Redefining Progress	3
Resources for the Future	0
Sierra Club	2
State PIRGs	0
Union of Concerned Scientists	0
Wilderness Society	1
Wildlife Society	1
WWF	1

[a] All websites for organizations were initially accessed on March 20, 2004. Coding was done at a later date.

occur in the same sentence in prominent, contemporary textual or programmatic material. These organizations are Center for Health, Environment and Justice, Center for a New American Dream, Environmental Defense, and Redefining Progress.

One intention of this chapter is to show that, despite the somewhat depressing overall picture, there are national environmental and sustainability membership organizations in the United States that are beginning to engage with the emergent JSP, according to my analysis. I believe their engagement is more than purely aspirational; it reflects their deeply held beliefs and values. Future research might characterize how these large organizations implement, frame, and shape their work and programs around just sustainability. In the next section, I will show how the JSP is being implemented today, in U.S. cities, primarily by small, local, community-responsive organizations.

Just Sustainability in Practice in U.S. Cities

To investigate further the JSP, as described in the previous chapter and as outlined in table 3.1, I want to turn now to a set of examples. These are *not* specifically programs or projects of the national membership organizations in my JSI survey (table 4.2), although they may have had some influence. Neither are they full case studies; rather, they are short, focused vignettes. I have simply put together a collection of five sustainability issue categories (land-use planning, solid waste, toxic chemical use, residential energy use, and transportation),[2] and three representative programs or projects per category, that are providing proactive, balanced efforts to create a just sustainability in practice in U.S. cities. The objective is to briefly describe a sample of local programs or projects, all of which have a JSI of 3 and illustrate different facets of the JSP in practice in the five issue categories.

The importance of these vignettes is that they illustrate that ideas of sustainability and environmental justice are being applied together, and in practice, in different locations and contexts. Many are based on multistakeholder partnerships between community nonprofits, national nonprofits, local or federal governments, and/or private industries. The avenues of implementation used at the community level are varied, involving tools and techniques ranging from the simplest and most reactive—street activism—through more deliberative processes and procedures typical of the JSP, to the most complex and proactive—building local economic security through private enterprise. While all have a JSI of 3, to show specifically how these projects or programs fit in the JSP, I have detailed their *main points of contact* after each vignette.

Most of the information about individual programs was acquired initially over the Internet in the summer of 2000, with some interviews with organization staff members. The information on each site was rechecked for accuracy, and some new information was added from December 2003 to July 2004. Although many of these organizations have been discussed in articles, either in academic journals or in the popular press, their websites were the most effective source of up-to-date information on their activities. For the most up-to-date information, I urge readers to visit the project or program website.[3]

Issue Category: Land-Use Planning

Historically, the primary tool of land-use planning, zoning, has led to geographic segregation of both people and land utility. Along with spatial segregation comes decreased social mobility. As Massey and Denton (1993:14) argue, "segregation constitutes a powerful impediment to black socioeconomic progress." Similarly, a recent report by the National Academy of Public Administration (NAPA 2003) found that when land-use planning and zoning laws are inadequate, the result is usually increased environmental and health hazards for communities of low income and of color. The report calls on local, state, and federal environmental, planning, and zoning agencies to launch environmental justice initiatives and to use them to solve existing problems and prevent future inequities (NAPA 2001a, 2001b). Clearly, zoning can lead to socioeconomic, environmental, and health disparities. Similar conclusions have been drawn by EJ activists and academics (Bullard 1994).

Planners concerned with sustainability point out flaws in land-use planning, such as separation of uses and low-density development, that have encouraged urban sprawl and auto-dependent transportation (Beatley and Manning 1997; Newman and Kenworthy 1999; Bullard et al. 2000; APA 2000). Recent movements in urban planning, such as New Urbanism and Smart Growth, together with their focus on transit-oriented development (TOD), however, have advocated for a change in historical land-use planning to encourage more efficient land development, mixed-use and mixed-income developments, and the reuse of former industrial sites (Duany et al. 2000). Additionally, procedural changes in the planning process now encourage greater community outreach and public participation in land-use decisions (Kelly and Becker 2000; APA 2000). However, as Hempel (1999) has argued, (just) sustainability in urban planning will require coordinated planning at the metropolitan/regional level,[4] while at the same time crafting collaborative and deliberative participatory approaches to comprehensive planning that prioritize existing community needs. In the meantime, community organizations, as we shall see below, are successfully developing tools to bridge the interests of their residents and the municipal planning process.

Representative Program 1: Urban Ecology, Oakland, California

Urban Ecology in Oakland, California, is an organization founded in 1975. As the website says,

> Urban Ecology has not focused on the traditional environmental priorities of preserving land, air and water. Neither have we had a traditional community development focus aimed at, for example, generating affordable housing. Rather, our work has integrated elements of these disciplines and others, with healthy "human habitats" as the common denominator. We have sought to advance sustainability in the Bay Area using three main strategies—alternative visioning, education and policy advocacy, with all of our work grounded in the three E's of environment, economy and social equity. (http://www.urbanecology.org)

Urban Ecology is engaged in two primary avenues toward promoting just sustainability principles in land-use planning within the San Francisco Bay Area. First, its Community Design Program provides planning and design services to low-income urban neighborhoods, such as the Weeks neighborhood in East Palo Alto, to assist them with community development. Urban Ecology has developed a process to bring the services of city planners into communities to engage in local needs assessments and community visioning. Urban Ecology helps organizations facilitate the drafting of a community plan that addresses the immediate and long-term needs of the area and assists the local community organizations with implementation strategies. Although the needs of the community are given first priority, Urban Ecology's staff promotes ideas such as transit access, pedestrian-friendly streetscapes, and affordable infill housing to help revitalize neighborhoods with sustainability principles in mind.

Second, Urban Ecology's Sustainable Cities Program approaches municipal governments such as Berkeley, Fremont, Oakland, and San Francisco and works with community groups such as San Jose's Tamien Neighborhood Association to promote more sustainable development patterns. The suburbs at the frontiers of urban sprawl are encouraged to adopt Smart Growth principles[5] that allow for diverse housing options and alternative transportation infrastructure. Urban Ecology advocates for infill development, affordable housing, transit-oriented development, reduced parking requirements, and mixed-use projects. It provides information to municipalities and citizen groups about private developers who have applied these principles in their projects. Urban Ecology also runs workshops for the public on how to review new projects and advocate for sustainable land development. In the Bay Area, the issues of urban sprawl, environmental preservation, and social justice are deeply linked together, and groups such as Urban Ecology are working with many communities in pursuit of more local and regional just sustainability.

JSI = 3. Main JSP points of contact: low-income urban neighborhood focused; use of deliberative community visioning tools.

Representative Program 2: Bethel New Life, Chicago, Illinois

Rioting and disinvestment in the late 1960s and early 1970s left this West Garfield Park community in Chicago in deep trouble. Bethel Lutheran Church members pledged to fight the despair, and in 1979 they bought a three-flat apartment building that became Bethel New Life. Now with 318 employees, 893 volunteers, more than 1,100 affordable housing units, 7,000 people in living-wage jobs, and $100 million invested in the community, this faith-based organization has gained, like DSNI in Boston, a national reputation for cutting-edge just sustainability initiatives.

The organization is a CDC whose strap-line, "Weaving together a healthier, sustainable community," reflects its wide-ranging asset-based community-development interests, which it pursues through programs such as cultural arts, employment, housing and economic development, family support, seniors, and community development. As its website states, "all programs & initiatives at Bethel New Life, Inc. are conceived with sustainability in mind, and must be: wanted by the community, financially viable and mission appropriate."

With regard to land-use planning, Bethel New Life's current major project is the Lake Pulaski Commercial Center. The project's team includes Farr Associates (architects), Phoenix Construction (contractor), Piper & Marbury (law firm), Matanky Realty (commercial leasing/operations), and Argonne National Laboratory (energy model and monitoring). The center is a 23,000-square-foot, two-story "smart green" building, a play on its "smart growth" and "green" qualities. With a bridge to the Lake Street El platform on the Green line, it is a TOD that will enable nonmotorized users quick access. Using photovoltaic cells, a "living green roof" that will enhance heat retention, superinsulation, and energy-efficient windows, as well as other energy efficiencies that combine to cut energy operating costs in half, it will house a child and infant daycare center, employment services, and five storefronts.

Major funding for this $4.5 million project comes from the City of Chicago Empowerment Zone, the State of Illinois Department of Commerce and Economic Opportunity, the City of Chicago Department of Environment, U.S. Bank, and Commonwealth Edison. A majority of the construction contracts are with Minority Business Enterprise/Women's Business Enterprise companies, which will create much-needed jobs in the community. In addition, almost seventy new permanent jobs will be created in food services, childcare, and retail. Another of the CDC's programs, Bethel Employment Services, will be housed in the center and will try to favor local community members in its recruitment drive.

JSI = 3. Main JSP points of contact: construction contracts are with Minority Business Enterprise/Women's Business Enterprise companies; use of local labor.

Representative Program 3: The Bronx Center Project—"Don't Move, Improve," New York City

The Bronx Center project has been recognized internationally through a UNESCO Management of Social Transformations program (MOST) Best Practice award and a 1996 Award of Excellence in Improving the Living Environment at the United Nations Conference on Human Settlements (Habitat II), held in Istanbul in June 1996.

The project is a collaborative, community-based revitalization plan. Focusing on a dilapidated three-hundred-block section of the South Bronx, it is a huge undertaking, requiring a multidisciplinary planning effort. It encompasses a wide range of different projects in economic development, health and human services, education and culture, housing, and transportation.

The center connects community members with urban development professionals, academics, not-for-profit organizations, cultural and social institutions, local businesses, and city officials and political representatives in an active and collaborative community problem-solving process. Organized with the help of the Bronx Community Forum, hundreds of frequently convened community forums and smaller working groups allow participants to discuss and find just and sustainable solutions to the social, economic, and physical problems of the neighborhood.

This is community-based participatory planning as advocated by the JSP. The planning process involves $2 billion in area revitalization activities over five years. It includes projects aimed at restoring buildings of architectural significance, like the Old Bronx Courthouse, to be reopened as the Bronx Planning Center; the construction of hundreds of new low- and mid-rise homes in Melrose Commons and the development of community-based health and human services facilities under the leadership of Nos Quedamos/We Stay; the refurbishment of present—and the development of new—learning and cultural institutions, such as the recently designed High School for Law, Government, and Justice to be housed in the proposed Supreme Court building; the creation of new green spaces and sports facilities; and the improvement of transportation systems. Perhaps most importantly, the Bronx Center, through its community labor exchange, heralds the creation of jobs and job-training programs so that residents can increase their earning potential and their dignity and expand their opportunities.

JSI = 3. Main JSP points of contact: community-based participatory planning.

Issue Category: Solid Waste Management

Solid waste reduction is one of the keystone issues of the NEP and the traditional environmental movement. The most widely practiced strategy, however, is recycling, although the hierarchy of actions should be "Refuse, Reuse, Recycle." Recycling is promoted as a municipal effort to reduce urban ecological footprints, partly because the public has seen it as "doing their bit," partly because it is heavily promoted by industry associations that do not want the public to move up the waste hierarchy by refusing or reusing their products, and partly because it is relatively easy to do if your municipality has a collection scheme.

At the same time, waste facility siting is one of the major issues confronted frequently by environmental justice groups (Cole and Foster 2001). To communities overburdened with waste management facilities, new projects involving trash, whether they are TSDFs or recycling facilities, are usually not a welcome land use. Sustainability advocates must use caution when proposing recycling-industry facilities as community economic-development opportunities for low-income areas. Waste facilities can be an *asset* in local economic development, contributing to work opportunities such as Garbage Reincarnation of Santa Rosa, California, but some waste facilities, primarily those for toxic waste, as was argued in chapter 1, can be an *assault* on such communities (Ackerman and Mirza 2001). Waste reduction, as opposed to waste management, must be planned so that there are environmental, social, and economic benefits and that the benefits are shared by all (Schnaiberg et al. 2001).

Representative Program 1: The Green Institute, Minneapolis, Minnesota

The Phillips community is one of the most diverse neighborhoods in Minneapolis, and it has a long history of community activism. In the words of the Green Institute,

> A fundamental aspect of our mission is the creation of high-quality living wage jobs for residents of the Phillips neighborhood, an area of concentrated poverty and unemployment. What sets us apart from many similar organizations is our emphasis on sustainable community development: development that simultaneously pursues economic, environmental, and social gains. (http://www.greeninstitute.org)

In the 1980s, the residents of Phillips organized an environmental justice campaign to resist the construction of a garbage transfer station in their community. The city cleared twenty-eight homes for the ten-acre site, but the construction of the project was eventually halted by the residents of the Phillips neighborhood. The People of Phillips neighborhood group then created the Green Institute to create sustainable business enterprises on the now-vacant site. The Green Institute is an entrepreneurial environmental organization creating jobs, improving the quality of life, and enhancing the urban environment in inner-city Minneapolis. It now operates three revenue-generating ventures designed to combine green industry with local economic development. First, in 1995, the ReUse Center was developed to sell scavenged building and construction materials. The retail store reclaims materials from the local waste stream and sells them at low cost. The center offers living wages for employees and offers community classes on home improvement. Second, in 1997 the Green Institute began a "DeConstruction" service to remove salvage materials from building or demolition sites. Through DeConstruction, up to 75 percent of an old structure can be reclaimed rather than demolished, with the materials sold at the ReUse Center. Third, the Phillips Eco-Enterprise Center, an award-winning business center built with green building technologies, was completed in 1999 on the site originally intended for the garbage transfer station. The Green Institute and its Phillips Eco-Enterprise Center are working to attract other environmentally conscious organizations and companies to continue their pursuit of sustainable economic development within the Phillips community.

One project that is in the pipeline is an urban energy cooperative that can help consumers deal with the challenges presented by the increasingly volatile, complicated, yet essential energy industry. The vision is to create a model locally and to provide national leadership for community households and businesses in their attempts to take control of the cost and conservation of energy. There are two complementary parts to this vision: to create a cooperative that delivers energy-related services through a demand-management model and to research the feasibility of a renewable biomass cogeneration facility designed to produce district heating and renewable electricity in the Phillips neighborhood.

In recognition of its efforts, the Green Institute was awarded the 1999 National Award for Environmental Sustainability from the PCSD and ReNew America. In addition, in April 2000, the Phillips Eco-Enterprise

Center was recognized by the American Institute of Architects as one of the Earth Day's Top Ten in exemplary sustainable design. It seems a shame that the 1999 award was not the National Award for Just Sustainability!

JSI = 3. Main JSP points of contact: combining green industry with local economic development in a diverse neighborhood.

Representative Program 2: The New York City Environmental Justice Alliance, New York City

The New York City Environmental Justice Alliance (NYCEJA) is a city-wide network that links community organizations, low-income communities, and communities of color in their struggles for justice. It was founded in 1991 to support community-based projects through a network of professional environmental advocates, attorneys, scientists, and health specialists. NYCEJA allocates resources to enable its members to be effective advocates for communities that are disproportionately and unjustly affected by the environmental and health impacts of public and private actions, policies, and plans.

In terms of solid waste activism, several communities are surrounded by heavy industrial areas, especially on their waterfronts. These areas have attracted private garbage transfer stations handling commercial waste from hotels, offices, and restaurants. These transfer stations bring in thousands of heavy diesel trucks each day. However, Fresh Kills landfill on Staten Island, the local destination for New York's garbage, was permanently closed in 2001, so the city has started sending some of its eleven thousand tons per day of residential garbage to these private facilities. This has nearly doubled the amount of garbage processed in EJ communities.

The Organization of Waterfront Neighborhoods (OWN), a city-wide coalition of groups fighting for just sustainability through their solid waste plan for New York City, was founded by NYCEJA in 1996. In "Taking Out the Trash: A New Direction for New York City's Waste" (Warren 2000), the aim is to maximize the sustainability—environmental, economic, and social—of the waste system by minimizing the export of waste and maximizing waste prevention and recycling. These options are cheaper, more environmentally sound, and can result in social benefits for low- and middle-income neighborhoods (cf. Ackerman and Mirza 2001).

Together, these groups have been successful in raising the profile of sustainability and have won significant legal battles to enforce NYC transfer-station siting regulations. At the same time, NYCEJA has helped organize Community Solid Waste Watch programs and developed a manual for local volunteers about the laws governing transfer stations and how to document violations. NYCEJA is also working on transportation justice issues at Melrose Station with the Bronx Center and Nos Quedamos/We Stay.

JSI = 3. Main JSP points of contact: proactive policy development: "Taking Out the Trash: A New Direction for New York City's Waste."

Representative Program 3: Reuse Development Organization, Baltimore, Maryland

The mission of the Reuse Development Organization (ReDO) is to promote reuse as an environmentally sound, socially beneficial, and economical means for managing surplus and discarded materials. Developed out of a conference in 1995 to fill a perceived information gap, reuse is the second priority in the solid waste management hierarchy after "refuse." Reuse means finding a use for something that someone thinks they no longer need. Although refusing something is preferable, Reuse is better than recycling, the third priority, because it conserves valuable natural resources, reduces the amount of water and air pollution, and reduces greenhouse gases, and it is a means for getting materials to disadvantaged people and organizations. Recycling actually uses a lot of energy. ReDO provides education, training, and technical assistance to start up and operate reuse programs.

As part of their Donations Program, ReDO has responded to many requests from nonprofit organizations and businesses that want an efficient, cost-effective way to give items that they no longer use to those who can use them. The program takes items that cost money to warehouse, transport, manage, and dispose of and provides a way of getting the materials to nonprofits that focus on people with low incomes, the ill, those assisting children, or the needy or disadvantaged. This gives businesses tax benefits (Internal Revenue Code, Section 170e3, "enhanced deduction") while building social capital in local communities. It also ensures that the donated materials stay out of the new-products marketplace.

JSI = 3. Main JSP points of contact: focusing profits from environmental industry on low income and underprivileged people.

Issue Category: Toxic Chemical Use

Four ideas have broadened the tools available to communities addressing the environmental justice and sustainability aspects of industrial operations. One is the *right to know* concept that requires full disclosure of chemical hazards to the community under the Emergency Planning and Community Right-to-Know Act (1986). This type of legislation is valuable, as a small-scale industrial operation using hazardous materials has the opportunity to create large-scale public health and long-term ecological risks. The second tool is *toxic use reduction,* which is aimed at redesigning industrial processes to use less hazardous substances and release less pollution into the air and water (Geiser 2001). Toxic use reduction allows for new production methods and the application of new technology rather than requiring plant closures. This functions as a tool against the "jobs blackmail" argument that industrial jobs in low-income communities will be sacrificed for environmental concerns. The third tool is the *Precautionary Principle,* which argues "where there are threats of serious or irreversible damage, lack of full scientific certainty should not be used as a reason for postponing measures to prevent environmental degradation" (Bergen Ministerial Declaration, quoted in Raffensperger and Tickner 1999:106). The sustainability and environmental justice movements would benefit from the creation of market demands that favor products generating less toxins and solid waste at the end of the line and from a regulatory system that enshrines the Precautionary Principle and promotes toxic use reduction. The fourth tool is *clean production,* which *Rachel's Environment and Health News* (1999:1) defines this way: "unlike 'pollution prevention' and 'recycling,' clean production asks fundamental questions about consumption: is a particular product even needed in the first place? And is it being produced in a way that promotes the goals of the community?" The first two tools are now well used by environmental justice and sustainability advocates, but the other two, the Precautionary Principle and clean production, are still relatively new and show huge promise.

Representative Program 1: The Silicon Valley Toxics Coalition, Silicon Valley, California

Silicon Valley has many claims to fame. Not only is it the hub of the United States' high-tech economic engine, but it is also the location of a high-tech toxic legacy and, not coincidentally, the birthplace of one of the high-tech industry's most diligent environmental watchdogs, the Silicon Valley Toxics Coalition (SVTC).

SVTC states its mission to be

> research, advocacy, and organizing to address human health and environmental problems caused by the rapid growth of the high-tech industry. Our goal is to advance environmental sustainability and clean production in the industry, as well as to improve health, promote justice, and ensure democratic decision-making for communities and workers affected by the high-tech revolution. (http://www.svtc.org/about/mission.htm)

Since SVTC's inception in 1982, the focus of the organization has been to raise awareness of the toxic health hazards, to workers and residents of Silicon Valley, produced by the superconductor industry. SVTC has sponsored key research on the location of toxic waste generating sites, areas of groundwater contamination, toxic solid waste generation and disposal, and the potential worker health threats of exposure to computer-industry processes in Silicon Valley. As part of its Health and Environmental Justice Program, SVTC honored a group of fifty immigrant women and men from low-income communities, at Olinder Community Center in San Jose, California, for their environmental justice work and commitment to their communities' health and well-being. This took place in April 2004, on People's Earth Day, an annual international event that emphasizes the need for people, especially those from marginalized communities, to reclaim the Earth and make it a healthier place for children and families.

Over the years, SVTC's focus has expanded both geographically and conceptually. In the globalized economy, manufacturing, research, trade, and waste disposal have all become international in scope, especially in the high-tech sector. SVTC has therefore begun working internationally with communities that are new sites of superconductor manufacturing. SVTC has worked with international standards in Europe for electronic waste disposal and has asked American-based companies to meet similar standards in the United States. This local-global linkage is characteristic of the JSP.

Like many environmental nonprofits, SVTC is initiating collaborative efforts with businesses to advance a just sustainability agenda. It has consulted with Silicon Valley–based companies on methods of toxics use reduction, improved environmental management, and solid waste recovery. It has also become a partner in a local government and industry sustainability-indicators project called the Silicon Valley Environmental Index. SVTC's work has thus engaged in a difficult balance of working with local companies while remaining environmental advocates for improved regulation and monitoring of their industry.

JSI = 3. Main JSP points of contact: local-global linkage.

Representative Program 2: Alaska Community Action on Toxics, Anchorage, Alaska

Alaska Community Action on Toxics (ACAT) is the only statewide organization focusing on environmental contamination and health issues. Its mission is

> to assure justice by advocating for environmental and community health. We believe that everyone has the right to clean air, clean water, and toxic-free food. (http://www.akaction.org)

ACAT works to empower communities to be involved in decision making and to ensure the responsible cleanup of contaminated sites. Their main focus, however, is on *toxic use reduction*: stopping the production, proliferation, and release of toxic chemicals. Their work is in four main categories: military toxics and health; northern contaminants and health; pesticide right-to-know; and water quality protection. Within these categories, ACAT's current major focus is investigating military abuses of the environment and health at installations and formerly used defense sites, including Amchitka Island (site of the world's largest underground nuclear test), Fort Greely, Gakona, Northeast Cape, Fort Richardson, Adak Naval Air Station, and military munitions ranges throughout Alaska.

In empowering communities, ACAT offers a range of services: Geographic Information System computer mapping (it has the only comprehensive database of contaminated sites in Alaska, with the mapping of two thousand military, oil and gas, mining, and other industrial sites); investigative research (it helps individuals and communities to find and interpret information, documents, and records through the Freedom of Information Act, the Internet, and literature reviews); advocacy (it works with EJ organizations around the country to prevent the production and proliferation of toxic and radioactive contaminants that threaten environmental and human health); and training (it facilitates meetings with scientific and medical experts, environmental justice and tribal leaders, organizers, and activists in order to share information).

JSI = 3. Main JSP points of contact: empowering communities to be involved in decision making.

Representative Program 3: Toxic Use Reduction Institute, Lowell, Massachusetts

Based in Lowell, Massachusetts, home of the United States' first manufacturing corporation, the Toxic Use Reduction Institute (TURI) is a university and state-office collaborative organization. It was created in 1989 under the state's Toxics Use Reduction Act (TURA) to decrease the quantity of toxic materials used and created by the state's industries. Indeed, TURI has helped industry reduce toxic chemicals used in manufacturing by 41 percent over the past decade, while improving the competitiveness of Massachusetts companies. Based within the University of Massachusetts, Lowell's School of Engineering, TURI researchers consult with companies and community groups working to reduce toxics use. The goal is to help industries continue production and contribute to local economic health, while cleaning up the environment in a state with a long history of polluting industrial practices.

The institute functions as the state's clearinghouse of resources on toxic use reduction. TURI conducts research on toxic use reduction technology, trains certified toxic use reduction planners, and distributes grant funding to cities, towns, and community or environmental organizations.

The grants are part of a toxic use reduction networking program that aims to develop model projects in Massachusetts communities. Some examples of its programs include healthier cosmetology practices, safer food production in school cafeterias, integrated pest management programs, and household hazardous products education.

Additionally, TURI funds and facilitates multiple public education programs regarding toxic chemical use. One important example of these programs was a two-day training workshop on clean production co-organized by the Lowell Center for Sustainable Production (a partner of TURI), the Deep South Center for Environmental Justice at Xavier University, and the Clean Production Network. Here, theoretical linkages and practical co-activism were explored, led by trainers from both the environmental justice and sustainability/clean production fields. Sessions included tools for clean production, lifecycle assessment, design for environment, sustainable product design, policies and resources for clean production, extended producer responsibility, ecological taxes, product lifecycle labeling, applying clean production in campaigns, brownfield redevelopment, and developing a vision for clean production.

JSI = 3. Main JSP points of contact: cooperative endeavor with the Deep South Center for Environmental Justice; clean production.

Issue Category: Residential Energy Use

Energy conservation in general is a win-win opportunity within the just sustainability agenda, as ACE is investigating with regard to energy-efficient affordable housing in Roxbury and as the Green Institute in Minneapolis is doing with regard to its proposed urban energy cooperative and renewable biomass cogeneration facility. Cutting energy costs can provide economic assistance to low-income residents, particularly in northern regions. Demand management with regard to energy resources has a long-distance benefit to communities affected by their proximity to mining operations, power plants, and hazardous waste disposal facilities.

However, the investment necessary to increase the environmental efficiency of existing homes and reduce the ecological impact of new home construction is often seen as incompatible with affordability goals. The result is that cities often rely on the "filtering principle" to generate affordable housing stock: namely, older, less energy-efficient houses become occupied by lower-income residents while wealthier residents purchase new houses. Older rental housing units create a particularly difficult problem in energy-efficiency policy, as the benefactor of home-infrastructure improvements is not always the owner. Even as new green building technology improves household energy efficiency, the challenge to broad energy use reduction will be creating the economic opportunity for technology investment and retrofitting old infrastructure.

Representative Program 1: National Center for Appropriate Technology, Butte, Montana

The National Center for Appropriate Technology (NCAT), established as a nonprofit corporation in 1976, works to find just solutions to environmental or economic challenges, solutions that use local resources and assist society's most disadvantaged citizens. It has developed multiple programs to address energy use for low-income communities. There are three noteworthy programs under its Sustainable Energy Program; the first two are current, and while the third has now ceased, it is mentioned because of the topicality of the issue.

First, the National Energy Affordability and Accessibility Project (NEAAP) is researching the impacts of energy-market restructuring and market changes on low- to moderate-income households. The project has a website and newsletter, and through the NEAAP Residential Energy Efficiency Database, domestic electric and natural gas customers can search for incentive programs offered by their local utility, such as home energy audits, energy-efficient appliance rebates, and loans at zero or low interest to upgrade insulation or replace old heating and cooling equipment.

Second, NCAT operates the Low-Income Home Energy Assistance Program (LIHEAP) as an information clearinghouse on residential energy conservation for those with the greatest energy cost burden and/or highest need. The program targets community groups, housing officials, energy providers, and low-income residents, providing information on conservation, energy self-sufficiency, and cooperative utility programs. The LIHEAP administers grants to help implement the goals of reducing the energy burden of households.

Third is the Affordable Sustainability Technical Assistance (ASTA) program that worked with Housing and Urban Development (HUD) grant programs. The goal was to incorporate green building designs into affordable housing projects in HOME-participating jurisdictions and subgrantees that work with the nation's neediest families. With a focus on outreach to Community Housing Development Organizations (CHDOs) in receipt of HOME funds, ASTA helped CHDOs that are largely understaffed and have limited in-house technical resources. Funding for this service was through HUD's SuperNOFA process.

JSI = 3. Main points of contact with JSP: multiple programs to address energy use for low-income communities.

Representative Program 2: Massachusetts Energy Consumer's Alliance, Boston, Massachusetts

Mass Energy, under its previous name—the Boston Oil Consumer's Alliance (BOCA)—was formed in 1982 to provide lower home oil heating costs through the buying power of bulk purchasing. With more than 7,000 residential members and 150 nonresidential members, Mass Energy collectively purchases more than five million gallons of oil per year, and with this enhanced buying power charges fifteen to thirty cents per gallon less than the average retail price, saving $150 to $300 per year per household.

Mass Energy's two-pronged approach is to increase both energy affordability and environmental sustainability. It does this through two community assistance programs: the Clean Energy for Communities Fund and the Oil Bank. The Clean Energy for Communities Fund is a new program aimed at supporting the installation of clean energy technologies at community-based nonprofits within its service territory. The Oil Bank program works each year through member donations that enable Mass Energy to help a small number of people who are put in the invidious position of choosing between food and heat. In 2003, it gave out more than $12,000 worth of heating oil to the neediest people.

In 2000, Mass Energy spearheaded the Solar Boston Initiative with a number of area nonprofits such as Episcopal Power and Light, DSNI, the Fenway Community Development Corporation, and the Tufts Climate Initiative, along with members of the solar energy industry. In partnership with the U.S. Department of Energy's Million Solar Roofs Program, the goal of Solar Boston is to serve as a link between the solar industry and consumers in order to reduce transaction costs of solar design and installations. Through consumer education, demonstration projects, and member consultations, Mass Energy has helped facilitate placing solar arrays on ten thousand homes in the Boston Area.

Following the state's recent deregulation, Mass Energy has also been developing a green electricity product, New England GreenStart, with options for members to purchase renewable electricity. The catch is that it is currently being offered only to Massachusetts Electric's (National Grid) 1.2 million customers in 168 Massachusetts communities. The state's major provider, NSTAR, does not yet allow its customers to purchase New England GreenStart.

JSI = 3. Main points of contact with JSP: Oil Bank and Clean Energy for Communities programs.

Representative Program 3: Communities for a Better Environment, Oakland, California

Communities for a Better Environment (CBE) is an environmental health and justice nonprofit organization that promotes

> clean air, clean water and the development of toxin-free communities. CBE's unique three-part strategy provides grassroots activism, environmental research and legal assistance within underserved urban communities. (http://www.cbecal.org/about/mission.shtml)

CBE currently runs Toxics, Oil Refineries, and Community Monitor campaigns. In addition, through its Power Plants Campaign, it has helped Californians learn about the state's highly publicized energy issues and organize against the Mirant Corporation–owned Potrero Plant. Mirant proposed expanding its plant in an already overburdened neighborhood of southeast San Francisco, which has two freeways and two major roads that carry a lot of trucks, resulting in poor air quality, high pollution levels, and health problems.

CBE argued that Potrero would produce an additional 625 tons of airborne pollutants per year for forty years, the life of the power plant. They continued that

> conservation, use of energy-efficient appliances and building practices, and increased use of renewable sources of power (wind, solar) will pro-

> vide California with the power to meet its needs, if the market manipulation by corporate energy producers is stopped. (http://www.cbecal.org/alerts/power/index.shtml)

The December 2002 "Electricity Resource Plan" by San Francisco's Environment Department and Public Utilities Commission supported CBE's conclusion about Potrero and marked the first government-proposed alternative to the Mirant Corporation's plan. The plan, argue CBE, would reduce local health risks because it

> would put 150 megawatts of mid-sized power plants in the city by 2004 while ramping up about 480 MW of electricity efficiency, solar, windpower, cogeneration, fuel cell, and other alternative technologies at many locations in and around the city by 2012. It seeks to phase out fossil fuel burning for the city's electricity over 20 to 30 years. (http://www.cbecal.org/alerts/power/pP0902.shtml)

JSI = 3. Main points of contact with JSP: low-income conservation vouchers.

Issue Category: Transportation Planning

Transportation justice, a response to discriminatory transportation planning, has addressed a wide range of issues over the past century, including bus and rail segregation, highway development, transit design, toxic freight, airport expansion, and neighborhood street safety (Forkenbrock and Schweitzer 1999; Conservation Law Foundation 1998; Bullard and Johnson 1997; Bullard et al. 2004). Historically, large-scale highway projects have had a significant impact on minority and low-income neighborhoods while facilitating increased automobile use and emissions by wealthier suburban residents (Bullard and Johnson 1997; Bullard et al. 2004). Activists are continuing to work to gain equity within transportation systems, particularly urban transit. In many cities, the difference in transit quality between services for suburban commuters and urban residents is analogous to the segregation fought in the bus boycotts of the 1950s and the Freedom Riders campaign in the 1960s (Bullard and Johnson 1997). Many just sustainability advocates point to transportation as the number-one issue to address in creating sustainable cities, and gradually federal, state, and local transportation agencies have included nonautomotive modes as relevant parts of transportation systems (Newman and Kenworthy 1999). One of the first steps in doing this, in controlling urban transportation futures for people, especially the disadvantaged, is to reframe the concept of transportation to the broader and more productive concept of *access*. By thinking in terms of access, we can think inclusively about the ways of bringing things to people and people to things.

Representative Program 1: Los Angeles Bus Riders Union, Los Angeles, California

Los Angeles is the premier example of the American automobile-dependent city. However, many people in the city's low-income communities are unable to afford car transportation and depend on the Metropolitan Transit Authority (MTA) for mobility and access. Ninety percent of the MTA's customers ride its aging and shrinking diesel bus fleet, while most of the agency's new capital investment has gone toward two underused rail lines. The MTA ran three thousand buses in 1984, but by 1999 its fleet was reduced to two thousand buses. Transit riders report that the MTA's buses are frequently overcrowded and breaking down. Meanwhile, the subway was running a $7 billion debt while attempting to attract suburban white commuters to its service (Rabin 1999a, 1999b).

The Los Angeles Bus Riders Union organized in the mid-nineties to fight an MTA fare increase, for what the union considered inadequate service. Through participation in public meetings and protests, the Bus Riders Union successfully prevented a fare hike that was intended to fund the construction of the subway. With the backing of an NAACP Legal Defense and Education Fund lawsuit, the Bus Riders Union has persuaded the MTA to comply with a court-ordered consent decree to purchase 248 natural-gas-powered buses in addition to the replacement of the existing 2,095 buses in the fleet. The union has also begun a Student Pass Campaign to simplify the process for students applying for student transit passes and to lower their cost. The Los Angeles Bus Riders Union has become a model for transit-riders organizers across the United States working to provide communities with clean, reliable, and affordable public transportation (Rabin 1999a, 1999b; Wexler 2000).

JSI = 3. Main points of contact with JSP: transit equity.

Representative Program 2: The Unity Council—Fruitvale Transit Village, Oakland, California

In the 1960s, a state agency was created to develop a unifying transit system in the San Francisco region, called Bay Area Rapid Transit (BART). As a transit system, BART has had mixed results and has come under a great deal of criticism for its high cost and focus on serving suburban commuter transit. An element of this commuter system design is that most BART stations include large surface parking lots.

When plans were announced for an expanded parking facility at the Fruitvale station in Oakland to serve driving commuters from outside the predominantly Latino neighborhood, the Fruitvale community responded with frustration. The Unity Council, a CDC for the Fruitvale neighborhood, developed an alternative plan for a TOD around the BART station. Through multiple community meetings and design charettes with assistance from the University of California at Berkeley, the community created a plan for a transit village at the location of the parking facility. Through rounds of negotiation, the Unity Council was able to convince BART and the City of Oakland to endorse its transit village plan, designed around pedestrian access to BART, retail development, and transit-oriented housing.

The mixed-use development uses ten acres of BART-owned surface parking and an additional fifteen surrounding acres. The master plan includes affordable housing, a senior center, a community health center, daycare facilities, street-level retail shops, and a hidden parking garage. The design incorporates streetscape elements and architecture that reflect the Latino heritage of the community. The transit village is the core of a neighborhood revitalization plan that also includes homeowners programs and local business improvement workshops to help existing residents benefit from new development. The community-based plan for a neighborhood center next to a transit station is an example of how innovations in transportation and land-use planning can meet the goals of just sustainability.

JSI = 3. Main points of contact with JSP: transit equity, community visioning.

Representative Program 3: Transportation Alternatives, New York City

Transportation Alternatives was founded in 1973, during the post–Earth Day 1970 explosion in environmental awareness that also produced the Clean Air and Clean Water acts and the EPA. Since then, Transportation Alternatives has been proactive in winning a wide range of improvements for the city's cyclists and pedestrians. It has also been the leading voice for car-use reduction. The organization's origins are in cycling, but it quickly realized that a cycle-friendly city requires a more just and sustainable transportation system.

Transportation Alternatives is looking to change the priorities of New York City's transit system so that it encourages and increases the more nonpolluting, quiet, and city-friendly travel modes and decreases —not bans—private car use. This policy shift, from *anti-car* to *anti-car usage,* has been achieved in many European countries and cities, where, despite high car-ownership levels, usage is far lower than in the United States because people walk and cycle more and because of low-cost, fast, safe, and efficient public transit alternatives. Transportation Alternatives is aiming for a rational transportation system based on a "green transportation hierarchy," which gives priority to travel modes based on their benefits and costs to society. The order of priority is pedestrians, bicycles, public transit, commercial vehicles and trucks, taxis, high-occupancy vehicles, and single-occupancy vehicles.

JSI = 3. Main points of contact with JSP: transit equity, auto deprioritization.

"Just Sustainability": From Theory to Practice

The JSI shows that there are a minority of national environmental and sustainability membership-based organizations in the United States that show a stated concern for equity and justice within the context of their work in environmental or sustainability issues. This is depressing, especially considering that most of the Big Ten, all of which are included in table 4.2, have not fully included a concern for equity and justice since the 1990 "Letter to the Big Ten" from environmental justice leaders. In fact, the only Big Ten organization that has a JSI of 3 is Environmental

Defense. Several international organizations mentioned, however, do have a JSI of 3, including the Worldwatch Institute, the Heinrich Boll Foundation, the Tellus Institute/Stockholm Environment Institute, the New Economics Foundation, and the Earth Council.

The more positive story is that all of the three representative programs or projects from each of the five sustainability issue categories of land-use planning, solid waste, toxic chemical use, residential energy use, and transportation represent a small sample of local, practical initiatives. They are demonstrating the implementation of the JSP in urban America. Perhaps it is because they are smaller organizations, not large national membership-based organizations, that they can be more locally responsive to the needs of diverse communities. Whatever the reason, these leading-edge projects show how inner urban communities can use asset-based approaches to develop their local economy in both a socially just and a sustainable manner.

In the next chapter, in order to look in greater detail at the JSP, I present a case study of Boston's ACE, an organization that illustrates, among other things, the interplay between just sustainability and environmental justice.

5

Alternatives for Community and Environment

In this chapter, I present an explanatory case study of the organization Alternatives for Community and Environment (ACE), based in Boston's Roxbury district. ACE was chosen as the unit of analysis because I believe that it, like all the other representative programs or projects described in chapter 4 with a JSI of 3, is demonstrating certain aspects of the JSP in practice in urban America. Although ACE identifies itself firmly with the environmental justice movement (ACE 2002), it is my belief that in its ten years of operation, it has changed both organizationally and programmatically. Is ACE still an environmental justice organization, or is its practice now that of just sustainability, or is it both?

In terms of programmatic change, which is my main focus, ACE seems to have changed in several ways. First, its programs have changed in scale, from predominantly local to more (metro Boston) regional. Second, its programs have shifted from predominantly single issue to more systemic. Third, its work is more coalition based, perhaps reflecting the complexity of systemic working and its reliance on wide-ranging skill sets, which can only be demonstrated by coalitions. Fourth, ACE's repertoire of action, composed of "institutional tactics such as lobbying, litigation, and educational campaigns or expressive tactics such as protest, boycotts, and street theatre" (Carmin and Balser 2002:366), is far more complex and nuanced than it was in the early 1990s when the organization was formed. These are broad generalizations and, as became clear through the interviews, there are exceptions. While none of these four changes per se indicates adherence to the JSP over the EJP, I think that the direction of change implicit in them is consistent with increasing adherence to the JSP.

Research Method

I chose a case-study approach because, following Yin (1994:13), "a case study is an empirical enquiry that investigates a contemporary phenomenon within its real life context, especially when the boundaries between phenomenon and context are not clearly evident." I carried out the research between January and June 2004, following a structured research design and protocol. I was fortunate in that ACE was in both a reflective and prospective mood, 2004 being its ten-year anniversary. Data was gathered from as wide a range of sources as possible, such as contemporary documents (including ACE's own *Alternatives Press,* program literature, reports, and scripts from a recent ten-year-anniversary video), archival records (including back issues of *Alternatives Press,* which it has produced since 1993, program literature, and reports), media texts (newspaper articles, op-eds), participant observation at Tuesday-morning staff meetings, and structured and semistructured interviews.

Following Yin (1994), I used these multiple sources of evidence, the triangulation of these sources, and the continual review by the ACE staff of the evidence and of my text, propositions, and conclusions to help ensure construct validity. I maintained internal validity by pattern matching, which Campbell (1975) describes as using several pieces of information from the same case to relate to a theoretical proposition. External validity was not an issue, as I do not intend to generalize based on one case; rather, my intent is to illustrate how one organization, ACE, relates to the JSP. I ensured reliability by developing a structured research design and protocol and a database of information.

In general, at staff meetings I took on the role of participant observer. In one meeting, however, I gained agreement on an ACE timeline (table 5.1), which, interestingly, the organization had never developed. I then used the technique of *critical moment analysis* to gain agreement on significant events in the organization's history. Critical moment analysis is the identification of key points in an organization's history, which are in effect "tipping points" in that organization's programs, activism, and even paradigm. Critical moments can be *internal* (new staff, resources, ideas) or *external* (local political windows, presidential executive orders, conferences, books, ideas). Critical moments are revealed by use of archival information and staff ideas and are more fully probed by the use of interviews.

TABLE 5.1
ACE Timeline

1993	
February 1993	Echoing Green Foundation awards $100,000 social entrepreneur grant to ACE.
August 1993	ACE begins first project, the Coalition Against the Asphalt Plant.
1994	
January 1994	ACE coordinates with Hands Across the River Coalition to stop a waste incinerator in New Bedford.
February 16, 1994	ACE incorporates as nonprofit in Massachusetts.
August 1, 1994	ACE opens Roxbury office.
November 17, 1994	Massachusetts Environmental Justice Assistance Network (MEJAN) brings lawyers and communities together.
1995	
February 1995	ACE coordinates with Hands Across the River Coalition to prevent filling Harbor Cove with PCB-laden material.
July 1995	Neighborhoods Against Urban Pollution (NAUP) collaboration develops resident solutions to existing environmental problems.
November 1995	ACE and DNSI reach agreement on illegal trash transfer stations.
1996	
May 1, 1996	ACE helps defeat asphalt plant. Working with the Coalition Against the Asphalt Plant, a coalition of residents from Roxbury, Dorchester, South Boston, and South End encourage the Boston Board of Health to vote against the asphalt plant.
January 17, 1996	First REEP town meeting at Orchard Park housing development.
June 20, 1996	REEP holds its first graduation.
June 24, 1996	ACE and the Chinatown Central Artery Task Force celebrate a victory over a proposed turnpike exit.
July 1, 1996	ACE moves to 2343 Washington Street.
1997	
March 19, 1997	First Jammin' for Justice.
July 21, 1997	First REEP Youth Summit.
October 22, 1997	Anti-Idling Day.
December 18, 1997	Clean Air for the Holidays.
1998	
1998	Clean Buses For Boston Coalition Fights for Clean Air.
March 5, 1998	Trash transfer station stopped in Lowell.
September 23, 1998	Breathe Out protest demands MBTA honor anti-idling commitments.

(continued)

TABLE 5.1 *(continued)*

1999	
April 10, 1999	Greater Boston Environmental Justice Network (GBEJN) launched.
September 26, 1999	ACE hosts the first Healthy Hair Show in New England.
November 5, 1999	ACE featured as one of Boston Phoenix's "Local Heroes."
November 16, 1999	Airbeat launched.
2000	
September 18, 2000	Fare increase rally. The T Riders Union rally against fare increase ends with meeting with Governor Cellucci.
October 26, 2000	T Riders Union wins bus-to-bus transfers and reduced-price weekly combo passes.
December 1, 2000	TRU launched.
2001	
Early 2001	ACE and its partners force MBTA to order 350 compressed-natural-gas buses.
July 18, 2001	Silver Line opens to protest.
August 10, 2001	No Air, No Fare rally.
Late 2001	ACE youth force cleanup of "mountain" of asbestos and lead-laden dirt in Roxbury.
2002	
March 2002	ACE and its partners win 100 clean-fuel buses in regional transportation plan.
July 2002	ACE youth bring more than 300 from Northeast together at 2002 Youth Summit.
October 2002	ACE and its partners push Massachusetts to enact its first Environmental Justice policy. After more than two years of pressure from ACE and its partners, Secretary of Environment Robert Durand passed the statewide EJ policy.
2003	
2003	ACE launches Beat the Fare Increase campaign. The On the Move Coalition and TRU organize transit riders to fight against yet another MBTA fare increase: "Why Pay More for Less?"
Spring 2003	Safety Net works to stop the BU Bioterrorism Lab.
August 2, 2003	T Riders Union leads "March on Washington Street" for transportation justice on the fiftieth anniversary of Martin Luther King's historic march.
2004	
April 1, 2004	ACE moves to 2181 Washington Street in Roxbury.
June 12, 2004	ACE celebrates tenth anniversary.

In essence, given my interest in the emergence of the JSP alongside the EJP and NEP, my research question was, How and why has the nature of ACE's programs and repertoires changed since it was founded? My propositions, based on some of the five differences between the JSP and EJP highlighted in chapter 4, are

P1: Organizations representative of the JSP show more *proactivity* than *reactivity* in their programs and repertoires toward developing sustainable communities.
P2: Organizations representative of the JSP use more *deliberative tools and techniques* in their programs and repertoires.
P3: ACE's programs are shifting from local and single focus to more regional and systemic.
P4: ACE is more likely to build coalitions with (other) organizations representative of the JSP than those representing an environmental sustainability orientation.

These propositions, which flow from the research question, formed the basis of my interview questions. The interviews were focused and semi-structured, with a core set of the same questions for staff[1] and a core set of the same questions for board members (and former board members).[2] The only "prop" in the interview was the ACE timeline (table 5.1), which allowed interviewees to remind themselves of the organization's history, programs, and critical moments. The semistructured nature of the interviews gave me the opportunity to probe each interviewee based on comments made in the interview, as well as, in the case of staff, what they considered critical moments and why. In all, fourteen interviews were carried out: two with the founders (and former board members Bill Shutkin and Charlie Lord), four with current board members (Bob Terrell, the chair, James Hoyte, Lisa Goodheart, and Gary Gill-Austern), and seven with current staff (Penn Loh, Warren Goldstein-Gelb, Jodi Sugerman-Brozan, Quita Sullivan, Eugene Benson, Khalida Smalls, and Klare Allen).

The Development of ACE

ACE now operates out of a newly rehabilitated office building at 2181 Washington Street that faces out onto the heart of Dudley Square in

Roxbury, one of fifteen neighborhoods in Boston. After establishing a growing presence in Roxbury over the past ten years, ACE is no longer a newcomer to the environmental justice scene, locally or nationally, but is a recognized player. From the beginning, ACE has been dedicated to working primarily within the Roxbury community to promote local empowerment in decision making for environmental, social, and economic issues.

The Roxbury community, southeast of downtown Boston, is 5 percent white, 63 percent black, 24 percent Hispanic, 1 percent Asian or Pacific Islander, less than 1 percent Native American, 3 percent other, and 4 percent multiracial, according to the U.S. Census (2000). The corresponding figures for the City of Boston are 50 percent white, 24 percent black, 14 percent Hispanic, 8 percent Asian or Pacific Islander, less than 1 percent Native American, 1 percent other, and 3 percent multiracial (U.S. Census 2000). In 2003, however, the city officially became a "majority-minority"[3] city, with people of color making up 50.5 percent of the population.

Alongside this demographic difference, Roxbury has higher percentages of people speaking community languages, particularly Spanish, French and French Creole, Portuguese and Portuguese Creole, and African languages, than does the City of Boston. In 1999, 73.2 percent of Roxbury's residents were classified as low to moderate income, compared to 56.2 percent for the City of Boston. These statistics reveal one of metro Boston's poorest neighborhoods, and a classic inner urban Environmental Justice Population, according to the Commonwealth's criteria as described in chapter 1. This is the racial, ethnic, and socioeconomic context in which ACE, a multiethnic organization, works.

ACE defines its mission in its 2002–2007 Strategic Plan:

> ACE builds the power of communities of color and lower income communities in New England to eradicate environmental racism and classism and achieve environmental justice. We believe that everyone has the right to a healthy environment and to be decision makers in issues affecting communities. (ACE 2002:1)

Although ACE initially began by marketing its legal and technical assistance tools to local Roxbury community groups, and to a few outliers in Vermont and in Lawrence, Massachusetts, it has now broadened its base to include the entire New England region, according to its mission.

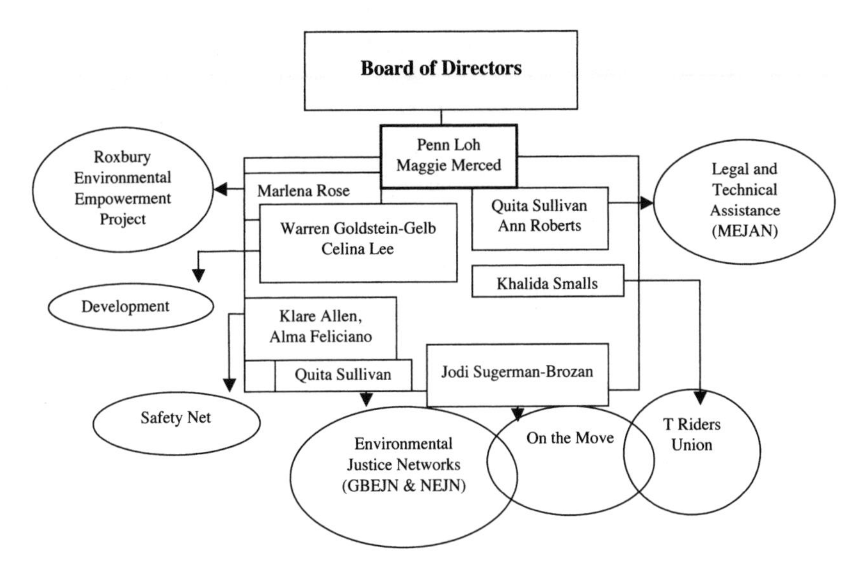

Fig. 5.1. ACE Organization and Program Structure.

To do this, it has sprouted many more programs of its own. These programs focus broadly on youth, transportation networks, and environmental health and typically involve cross-boundary coalitions, which are currently managed by members of ACE's staff of fourteen (see fig. 5.1), although this will likely change as ACE's membership plan unfolds. In these programs, ACE is supported by hundreds of activists in Roxbury and communities primarily across metro Boston and eastern Massachusetts. In short, ACE is now a strong part of Roxbury's history, helping to build more power within communities fighting EJ issues in metro Boston and the larger New England region.

Theory of Change and Model of Community Action

ACE is an organization that lives by its principles in all aspects of its work, be it coalition building, EJ movement building, its program selection criteria, or its "family building," the nurturing of talent within its staff and local community. Particularly noteworthy in this family building is its nurturing of local youth. Stanley Wiggins and Nwando Obele, former ACE youth interns and now Brandeis College and Mount

Holyoke students, respectively, are both board members. David Noiles, another former ACE youth intern, now works part-time on the Roxbury Environmental Empowerment Project (REEP) and attends Springfield College. David leads ACE's "toxic tours" around the neighborhood, and my students and I have benefited from the insights to the community that this young man has given over the years. In addition to youth, ACE employs many people from the local community including Klare Allen, Khalida Smalls, Marlena Rose, Maggie Merced, Alma Feliciano, and Quita Sullivan. For a not-for-profit on the usual funding rollercoaster, ACE has had a remarkably stable staff. I believe this is in large part because its staff share its principles, especially its family-building role.

In terms of principles, two phrases came up again and again as ideas or concepts that underpin ACE's entire enterprise: *popular education* and *empowerment-practice,* which according to Warren Goldstein-Gelb is "the analogous piece to popular education on the legal side." The classic guide to popular education is Paulo Freire's *Pedagogy of the Oppressed.* This and later books and articles sought to build community capacity for democratic social change through education. Proponents of popular education believe that, fundamentally, the purpose of education is social transformation toward full human participation in society. Popular education came to ACE first through its youth programs and then spread throughout the organization's programs. Worth mentioning here is that the empirical companion to popular education is participatory research, mentioned as CBPR in chapter 1. ACE is very cautious about research, which it sees as simply experts coming in, gathering data, and then leaving, with no tangible community benefit. While the case-study approach I adopted clearly involves many people, our "added value" to ACE was to create a display-size timeline, complete with photos, of ACE's history, which was unveiled at ACE's ten-year-anniversary party on June 12, 2004.

Empowerment-practice works on a strengths and assets perspective, centering the process of building community power on developing individual and collective competencies. This approach operates on the basis that whole communities benefit when they, and organizations like ACE, acknowledge every person's rights and responsibilities to contribute to and receive from community participation, in a relationship with reciprocal obligations.

ACE's theory of change flows from this idea. Bill Shutkin explains that

> part of ACE's theory of change was that we were bringing to communities the same kinds of resources that wealthy, more affluent, whiter communities possessed all along, and in that sense we were at least leveling the playing field. But we understood as a matter of our theory of change that leveling the playing field wasn't enough. We weren't just about everyone being able to fight against unwanted development. We were about just sustainability. We were about the idea that all communities, especially EJ-type communities should have the ability, the opportunity, capacity, and wherewithal to chart their own destiny, what kind of land use, what kind of jobs, what kind of skills, what kind of housing, etc.

Here, Shutkin makes a similar point to mine in chapter 1 that improvements to right environmental injustices in EJ communities should go further than just "bringing them up to scratch" (i.e., up to our current state of *unsustainability*). He also makes two other very significant, related points about the family-building aspect of ACE. He notes that

> part of our theory of change too was that we would staff up ACE with the very kinds of folks we were working with in the community so that internally and externally we were reflecting the same message of empowerment, of capacity building individually and at the community scale

and that "we would bring in that rough edge combined with real professional training, capacities, and competencies into ACE." These last two statements proved to be very important in explaining the internal dynamics of the organization, which helped create an external dynamic important in a community, Roxbury, that is multilingual and predominantly of low income and of color. On getting to know ACE over the six months of my case study, it became apparent that this is an organization of immense principle and virtue, yet it is also a very pragmatic organization. The former became clear through examining the coalitions ACE forms or, more correctly, who they will and who they will not work with and why. The latter became clear through looking at the individuals involved and their roles. Shutkin's explanation of ACE's theory of change seems to be working. The organizers, the ones out in the community, are largely a reflection of the demographics of the Roxbury community—African American (Khalida Smalls, Klare Allen, Marlena Rose), Latino (Alma Feliciano), and white (Jodi Sugerman-Brozan)—while the ones in the office are more white and professionally qualified with bachelor's or master's degrees (Penn Loh, Asian American; Warren Goldstein-Gelb, white; Eugene Benson, white; and Quita Sullivan, a member of the Montaukett tribe of Long Island, New York). Maggie Merced, the bookkeeper, is Latina, and Celina Lee, the development assistant, is Asian American.

This division did not go unnoticed among both board members and staff. A quote, which I agreed not to attribute, regarding the first African American employee, Klare Allen, made it clear that this organizational set-up was no accident:

> Klare Allen on her own merits is a compelling figure and was the day we first met her and I first met her. From an instrumentalist point of view, we needed a Klare Allen. We had Joanne Henry, a [black] Berkeley-trained lawyer who was dating white guys, but we needed the real deal, authenticity, and that's the instrumentalist in me talking.

Klare Allen is clearly a pivotal figure in ACE's success. Before joining ACE, she founded the Mothers Coalition, a group dedicated to promoting the interests of homeless women and their children, a situation that she had experienced. She has been given many awards for her community organizing and her teaching. In 1999, she was honored with the *Parents* magazine As They Grow Award. In 1998, she was presented with the African Achievers Award from the Black Community Center. In November 1996, she was Conservation Teacher of the Year, an award bestowed by the Massachusetts Audubon Society for her role in launching the REEP. Earlier that year, she also received the Green Leaf Award from the now-defunct Environmental Diversity Forum, set up in the early 1990s, by current ACE board member James Hoyte, as a coalition of individuals, community groups, environmental organizations, agencies, businesses, and grant makers working to diversify the environmental movement.

Allen's colleague and friend Khalida Smalls says, "a lot of the way Klare is, the type of person that Klare is she can't . . . she doesn't remember anybody's freakin' name, but she knows who they are, you know what I mean, and they know who she is." She continues, reflecting on the division within ACE and the role she shares with Allen, by saying she thought it just "panned out that way." However, she thinks that now "it has come to the point where it's beginning to be a conscious thing," not just to her and Allen but also to people outside the organization. Smalls does not appear to have a problem with this:

> we joke about it all the time, and we're like, "we're the black representatives, you need a black representative" . . . but it's something, at least for me, that has only really become a clear thing in the past year, two years, whatever, but it wasn't like an intentional thing, it was just kind of how it ended up. I think it is one of the things that make ACE credible, that makes other people want to be part of the organization or find out more about what the heck it is doing.

Bob Terrell, an African American who is director of the Washington Street Corridor Coalition (WSCC), a community-based transit-advocacy project in Roxbury, and ACE's chair, talks about the division in ACE as "uneven development" in the organization, which he did not necessarily see as college-educated versus non-college-educated:

> because some of the college folks knew certain things but didn't know other things and the younger activists out of the black community that worked there [at ACE] knew certain things locally but didn't know other things historically, . . . there was all kinds of uneven development amongst the entire staff. But what I liked about ACE was there were young folks; they were working very hard, very dedicated, very disciplined.

Like many others I interviewed, James Hoyte, an African American who is assistant to the president of Harvard University and a long-time ACE board member,[4] sees the flip side of the coin, that mutual respect is the key to ACE's success:

> I don't know. It's worked. You often scratch your head in a sense and say in a way it's amazing that it's worked. It's to do with the particular values and skills and the personalities involved. Underlying all that is this kind of mutual respect that all those folks have. I think if you didn't have that there, it could almost seem like, not apartheid, but division.

Klare Allen makes a crucial point about ACE's work that goes a long way toward explaining why it is that the organization's people of color

TABLE 5.2
ACE Revenues over the Last Five Fiscal Years

	Carryover	Foundation	Gov't & Contracts	Individuals	Other	Total
FY99–00	$89,465	$367,693	$170,539	$47,934	$6,888	$682,518
FY00–01	$104,586	$572,900	$99,564	$67,809	$13,302	$858,161
FY01–02	$101,125	$491,479	$31,875	$85,122	$17,841	$757,367
FY02–03	$36,824	$717,708	$20,163	$100,322	$38,778	$913,794
FY03–04	$27,975	$747,854	$8,750	$81,060	$2,656	$862,984

are the primary communicators with the local community. She says, "it's not the message you're putting out there but who's bringing that message." The implications of this statement are profound, not just for ACE but for all organizations working with diverse constituencies, not for profit or otherwise.

With regard to funding, ACE has grown significantly since receiving its first two-year grant from the Echoing Green Foundation in 1993. The total revenues and diversity of revenue sources has increased as ACE has increased its programs over the years. The total number of full-time salaried staff rose from the original two to its current peak of eleven. Like many nonprofits, ACE makes the most of its precious dollars by leveraging its interns and volunteers to perform many of ACE's duties. The near future will also see ACE asking its members for monetary contributions as well as time contributions. This movement toward membership is intended to offset the often volatile mood swings of foundation grants. The past five years of ACE funding reflect the economic market fluctuations (see table 5.2).

In terms of a model of community action, Penn Loh, ACE's executive director, who joined ACE in 1995 and has served as research and development director and more recently as associate director, critiques the Saul Alinsky "Rules for Radicals"–style of organizing, which he sees as rather superficial: bringing people together to define a winnable goal, winning some local victories that may be of no consequence in the systemic or bigger picture, without fundamentally changing people's consciousness as to why there are fundamental inequities in society in the first place.

The ACE environmental justice model is, Loh argues,

> not about surface solutions. It's always been about finding deep, systemic solutions, which means that if the movement wants to be grassroots led,

really community driven, you can't do anything but try to work with people so that they are building their own perspective, so the tools that we have are around "how do you do that?" and "how do you do that and help folks develop strategies that do end up with things that they can see as progress?" and that's the continual challenge. I wouldn't say that we've come up with the right formula. We've tried different things.

Programs

ACE currently has five primary program areas.

1. Roxbury Environmental Empowerment Project

REEP focuses on the development of youth environmental justice leadership in Roxbury neighborhoods through ACE's environmental justice curriculum, internship program, and youth-led projects.

2. Safety Net

ACE empowers residents in public housing and others in Roxbury to develop a voice and vision of a "Sustainable Roxbury" and equitable metropolitan development.

3. Environmental Justice Networks

ACE staffs the Greater Boston Environmental Justice Network (GBEJN), which brings together thirty neighborhood groups. It is also the New England coordinator for the Northeast Environmental Justice Network (NEJN).

4. Transportation Justice

ACE is the facilitator for On the Move: Greater Boston Transportation Justice Coalition and is the founder and home of the T Riders Union (TRU), a group with more than five hundred members that organizes transit riders. Both groups focus on transportation to improve the environment and quality of life for low-income neighborhoods and neighborhoods of color like Roxbury.

5. Legal and Technical Assistance/Services to Allies

Working in close partnership with community groups in greater Boston, ACE attorneys provide legal representation combined with community capacity building and organizing assistance. ACE also coordinates the Massachusetts Environmental Justice Assistance Network (MEJAN), which has grown to a network of more than one hundred attorneys, public health professionals, and environmental consultants who provide pro bono assistance to groups throughout the state.

Program Selection Criteria

To achieve its mission and in keeping with its principles, ACE has developed a set of criteria for selecting programs. Past ACE campaigns derived from REEP-sponsored direction, Roxbury residents, and allied community groups. However, limited resources and an overwhelming demand for community support now place ACE in a role of constant prioritization of new programs. In order to create the systemic changes it sees as vital, and to sustain power in the region, ACE compiled a three-tier system for program assessment and, ultimately, selection that is based on its foundational principles, popular education and empowerment-practice.

Tier 1, Criteria for Meeting ACE Mission, calls for programs that address priority concerns derived by, achieved for, and led by the Roxbury and Greater Boston residents. This level empowers residents by seeking their active input at all levels of decision making and implementation.

Tier 2, Criteria for an Effective Campaign, calls for easily understood goals that can be achieved within a clear time frame. The program should also provide the foundation for future community partnerships, increase resident awareness of EJ issues, and allow future campaign development.

Tier 3, Criteria for Sustaining ACE as an Organization, builds on ACE's strengths, allows fundraising support, and creates unique programs that can be shared across the country. This last element ensures that ACE remains a leader in developing new programs and continues to provide support in the form of replicable models to other organizations nationwide (ACE 2002: Attachment A).

Programs, Repertoires, and Communication

Each of ACE's programs operates under the ACE logo but currently retains a certain degree of autonomy as an independent campaign. This allows ACE to coordinate programs within its overall mission without creating a rigid organizational structure that is unresponsive to community desires and inputs. This will change, however, as ACE develops its membership scheme. Each program is different in its membership, repertoire, and communication techniques, with the program coordinator or director/co-director often determining the preferred repertoire(s) and means of communication. Programmatic report-backs or updates to all staff are primarily through the mechanism of the weekly staff meeting.

In the case of REEP (co-directed by Klare Allen and Jodi Sugerman-Brozan and coordinated by Marlena Rose), most work is done in person and through teach-ins and local meetings with students, interns, community members, and ACE staff. The REEP program recently started issuing a newspaper-style annual report similar to ACE's *Alternatives Press* newsletter.

The Safety Net program (organized by Klare Allen, with assistance from Alma Feliciano) also primarily uses in-person meetings and teach-ins and word of mouth among residents to spread the news about recent events. At the same time, pamphlets and news media coverage of the program's achievements also assist in increasing residents' awareness. Klare Allen maintains an extensive network of concerned residents that she communicates with primarily verbally to rally support for projects—including the focus of current Safety Net work, the $128 million Boston University Level-4 biolab proposal. A wealth of rebuttal information to Boston University's statements as well as general information on Level-4 biolab facilities can also be found on the ACE website. This campaign is one of the first to make increased use of the ACE website (http://www.ace-ej.org/BiolabWeb/Biolab.html), although ACE plans to make great strides to bring more campaign information to the Internet in the next year.

The On the Move Coalition (coordinated by Jodi Sugerman-Brozan) and the T Riders Union (coordinated by Khalida Smalls in partnership with Jodi Sugerman-Brozan) work within their own constituencies to educate and empower residents in dealing with transportation issues. Their educational approach consists of teach-ins, whereby a group of TRU members will meet locally to discuss Massachusetts Bay Transportation Authority (MBTA) issues and learn more about the region-wide transportation system. Members often feel validated to know that their concerns about MBTA issues are shared by others in the community and across the region. Pamphlets and workbooks explain the general regional transportation system and then encourage residents to take action in addressing their primary concerns.

The Legal and Technical Assistance/Services to Allies program, currently focused on MEJAN (coordinated by Quita Sullivan and volunteer Anne Roberts), also works primarily within its own constituencies to keep lawyers and communities aware of new developments in EJ communities. Since each case is different, MEJAN often assumes a one-on-one approach to remain in contact with pro bono lawyers working out specific case issues in a community. As court victories arise, there are announcements in the local media and *Alternatives Press*.

The Environmental Justice Networks, currently GBEJN and NEJN (GBEJN is coordinated by Quita Sullivan, and she is secretary of NEJN), are a key aspect of ACE's movement-building work. GBEJN organizes the annual "EJ in the Hood," a teach-in-based conference for local activists and those in local government, federal agencies, and advisory bodies such as the Boston Metropolitan Planning Organization.

Roots and Shoots

The early roots of ACE extend back to the 1990 National Environmental Law Students Conference that brought founders—and, until recently, board members—Bill Shutkin and Charlie Lord together in New Orleans. During a speech, a group of activists stood up and interrupted the speaker, an advocate of the NEP, the traditional or environmental sustainability movement. The activists, Native Americans, African Americans, and white women, said that their communities in the South were suffering from toxics due to policies that this particular NEP group and others like it advocated. They argued forcefully that they needed lawyers to defend their communities and that they also needed national environmental groups to focus on these community impacts. Unfortunately for the activists, the speaker was from a group with a JSI of 0, so they were barking up the wrong tree.

However, the outburst from the crowd at the speaker's lack of understanding of local EJ issues created in Shutkin and Lord a sense of urgency that these issues must be addressed, and in many more communities. It was ACE's first "critical moment," around the time of the "Letter to the Big Ten" from EJ leaders. The combination of what they had seen at the conference and the increasingly uneasy relationship between the traditional or environmental sustainability movement and the flowering EJ movement could not have been more charged.

The result of these confrontations is now history. Instead of working with the government or established nonprofits, Shutkin, who was working on a PhD at Berkeley, and Lord, who was clerking for a judge in Washington, D.C., developed the idea of an *entrepreneurial advocacy organization.* Given what they had recently been through at the conference, Lord's rationale for this embryonic organization seems rather understated: "I'd always felt that the environmental movement as it had presented itself to me as a place of work and vocation was missing an element of social justice which meant a lot to me." After receiving $100,000 for two years of funding from the Echoing Green Foundation and free office space from Boston College Law School, ACE was open for business. Shutkin remembers the funding: "the foundation that was new at the time . . . was willing to invest in two white guys claiming to want to do work in the inner city with people of color communities and claiming to want to essentially join this emerging EJ movement." Neither one had moved in before they started making phone calls to contacts made during the previous year. Right away, there was a call-back about an asphalt plant that needed a lot of work, and quickly. It was an interesting first case about ACE's core issues of community power and environmental legal support. ACE was also fortunate enough to have partners, such as DSNI, which were willing to work with it, and this first case provided the opportunity to strengthen bonds. As the director of DSNI at the time, Greg Watson, said,

> I can honestly say that because of Charlie Lord and Bill Shutkin, ACE really is considered a partner in our effort to revitalize our community. And they have done it in ways that have not been pretentious. They have redefined what it means to be lawyers working with groups like ours. They came in not as experts. They wanted to share our vision. (*Alternatives Press* 1998:7)

This accolade is typical of the many that I heard during my case-study research period.

The First Case

In the year before ACE was established, Tedesco Equipment proposed building a sizable asphalt-batching plant in the South Bay area of Boston, near I-93 on a 2.5-acre parcel near the Boston Municipal Incinerator and next to the entrance to Lower Roxbury. ACE was brought into the case by Stephanie Pollock, senior lawyer at Boston's Conservation Law Foundation (CLF), an organization with which ACE was to have an uneasy relationship in the future. There were a few residents organizing about this NIMBY issue already, and ACE immediately began getting up to speed on the case. The Boston Zoning Board of Appeals approved a permit for siting of this facility in the summer of 1993. ACE had only two weeks to bring a suit in Massachusetts Superior Court against the permit. The suit originally focused on the impacts to local residents based on air quality and traffic. ACE eventually constructed an argument, one of the first in the country, that linked small amounts of air pollution to asthma.

Then, two weeks later, Harvard University released the famous "Six City Study" demonstrating that particulate air quality was linked to respiratory problems (Dockery et al. 1993). Within a year, the asphalt plant became a different case: it was transformed into a demonstration of harm to local residents who were nonabutters. Roxbury residents already suffered from the highest levels of asthma and upper-respiratory disease in Massachusetts (Levy et al. 2001). ACE tried to frame the case as a new public-health challenge that presented the question, "What is the cumulative effect of these numerous small sources?" Each of these small sources was under the threshold of harm, but together they represented a serious threat. The local board of health could now act beyond the constraints of regulatory thresholds and apply this new public-health data.

The asphalt case was the first of its kind to link small air quality polluters with asthma and upper-respiratory disease. Shutkin recollects,

> clearly the Asphalt plant case, as really our first project, though a litigation, but a project nonetheless, got us on the map. It was the opportunity that allowed us very quickly, instantaneously, to arrive on the pages of the *Boston Globe,* to become known to a large constituency in the heart of Boston across geographies, across communities, and it allowed us in a sense to ply our wares.

The Massachusetts Environmental Justice Assistance Network

The Massachusetts Environmental Justice Assistance Network (MEJAN) involves planners, lawyers, and other environmental professionals in community efforts. It offers professionals a chance to do community work. MEJAN averages about seventeen to twenty cases a year throughout the metro Boston area, and the cases range from landfills in Lowell to industrial permitting in East Freetown. Developers come to the table with scientists, lawyers, and planners, whom ACE coordinates on behalf of communities. The preferred outcome is not always a court victory but can be a compromise amenable to the local community.

MEJAN is an outgrowth of the technical-assistance work that ACE originally started locally in 1993. MEJAN works with communities of color and low-income communities, which often requires lawyers to have special skills and cultural understanding. The collective community memory is very long and community awareness is high, so MEJAN lawyers must be attuned to many issues. Prior to uniting participating lawyers with community groups, ACE spends time coaching the professionals on dealing specifically with communities of color and low-income communities. Once connected, MEJAN planners, lawyers, and other environmental professionals spend time interpreting master plans, challenging permits, and being available as legal resources for communities.

The program is not entirely focused on reactivity, or stopping "bad" projects. Increasingly more important than simply stopping a project is proactivity, or deciding what the community wants and creating opportunities for the empowerment of local residents. ACE seeks to bring development that will benefit the community and has begun to work on forward-looking issues, such as the Roxbury Master Plan. Other communities focus on principles to let developers know what the communities do and do not want to see in future developments. These MEJAN-assisted planning processes allow for incoming developers to create better projects that are more acceptable to the community.

Roxbury Environmental Empowerment Project

One of the original ACE principles, in addition to technical assistance to local communities, was a program dedicated to youth empowerment. The REEP started with the idea of bringing "the environment" to middle school and high school students. Rather than busing them out to someone else's environment, the popular-education model seeks the development of students' awareness, skills, and confidence so that they want to work for the betterment of their own environment. The model also sees developing such skills as transferable; once learned, they can be applied in other instances, issues, and situations. In 1996, ACE developed a popular-education-based environmental curriculum related to community issues, so that youths could better visualize the issues in their neighborhood. Of the fifty schools invited to participate in this new curriculum, only five schools accepted ACE's offer to lead a new course. REEP has since worked in many different classrooms and with many different age groups and has become a national model for nurturing youth leadership in the environmental justice movement.

After several classes, many youths wanted to continue their involvement in these issues, and the after-school program started from these first engaged students. The students would soon take ownership of a wide range of self-selected projects—from collecting guns to promoting healthy hair products. ACE advanced the notion that student's ideas were never incorrect, and it tried to encourage the students in all their efforts. This epitomizes ACE's commitment to youth by actively listening and responding to them rather than simply projecting adult ideas into the REEP program. In fact, one early REEP focus on asthma evolved into the largest ACE program on transportation and air quality.

Youth really brought the asthma and environment issue to ACE and let it be known that this is an extremely important issue in the community. The REEP program was very flexible in allowing young people to pick the topics of discussion. The youths picked asthma as their topic when one youth in a school suddenly died, and it became evident among the students that asthma was a serious concern. Air quality issues were already prevalent in Roxbury, since many youths would close windows in the summertime, continuously clean windowsills, and stay indoors during the summer to avoid the fumes. As the Massachusetts Senate Committee on Post Audit and Oversight (2002) said,

> Urban areas are particularly hard hit with higher asthma rates. In Roxbury, a neighborhood of Boston, State Senator Diane Wilkerson visited an afterschool program and noticed asthma inhalers in 17 out of 20 cubbyholes. In Roxbury, the rate for asthma-related hospitalizations among children is five times the statewide average.

This was one of the motivations for youths taking action in the Anti-Idling March, another ACE "critical moment." When ACE first started doing this work in 1995, it was commonly known that asthma was a medical problem, but not much else was certain. The REEP youths began to find out and promote the idea that asthma was an environmental health issue too. Part of their solution was good health care, but another preventative solution was better air quality. There are many groups working on asthma today—doctors, hospitals, and schools—that recognize that asthma is both a public health issue and an environmental health issue.

However, the issue took a much more public face in Roxbury on October 22, 1997, when seventy-five youths from Nathan Hale Elementary School and Greater Egleston Community High School proposed and organized a march to publicize the Massachusetts anti-idling law. According to the EPA (2004), this is "a Massachusetts law and regulation to prohibit unnecessary idling of all motor vehicles that are stopped for a foreseeable period of time over five minutes" (chap. 90, sec. 16A, and 310 CMR 7.11). Jodi Sugerman-Brozan, who had until recently worked for Save the Harbor/Save the Bay, a group in the traditional environmental sustainability movement in Boston, recalls the day vividly:

> I remember my first class. I started in September, and on October 22, 1997, was the big Anti-Idling march. And that was my very first project, the very first thing I went on. And I remember marching down the street with these kids rallying, calling for clean air. And I pinched myself and said, "this is my job, I get paid for this." It was the best.

In the months before the Anti-Idling March, ACE's repertoire of action was dominated by community-mapping projects in the neighborhood in order to find all the sources of air pollution (Levy et al. 2001). The diesel buses and trucks in the area were numerous, and many of the MBTA bus depots were located in the area. According to *Alternatives Press* (1997:1), "within 1¾ mile of Dudley Square are more than 14 bus and truck depots housing more than 1,150 diesel vehicles, including 500 public buses and 230 school buses." Most residents felt that all these diesel buses could not be good for air quality, but there was a need to get more scientific data. However, younger people invoked what was essentially the Precautionary Principle. In their eyes, the connection between diesel fumes and particulates and air quality was self-evident, and in the absence of full scientific proof, they decided to launch into action. These same youths lived across the street from bus depots and knew that the buses idled for longer than five minutes at a time and, in so doing, broke the state's anti-idling law. In fact, the youths were saying that they kept their windows closed because the buses idled for up to an hour before going out on the road. Again according to *Alternatives Press* (1996:7),

> after students discovered that Roxbury had shockingly higher rates of asthma compared to the rest of Boston, they began a project to identify the causes of asthma in their neighborhoods. The problem was confirmed not only by studies done by public health professionals, but by their fellow classmates—everyone in their class either had asthma or had a relative or friend that suffered from asthma.

ACE brought in environmental officials from the city to listen, to see (and smell) these issues firsthand. The city officials committed to work with the youths on this issue. At first, the youths wanted to give fines directly to the bus drivers, but eventually their plan evolved into giving drivers an "educational ticket" that gave information about idling. The Anti-Idling March actually involved students wearing respirators and handing out these tickets to educate people about the effects of idling and cars. They designed this event so that the REEP youths were themselves the key speakers and organizers. EPA officials were on hand only to congratulate the youths on their work. The story broke on television that day when Nathan Hale Elementary School students as young as ten were coming forward on the regional news channels to tell the story of how these air quality issues were affecting them.

The immediate result of the Anti-Idling day was a commitment from the MBTA to stop bus idling and to inform drivers about not idling. However, the MBTA only undertook minimal compliance with this issue by posting several signs and issuing a few memos to drivers. In 2000, the EPA stepped in to begin an investigation of the problem. The EPA videotaped fifty-six MBTA buses idling between February 7 and February 21, 2002—for up to two and a half hours at one Roxbury station. An EPA decision announced on March 10, 2004, called for $1.3 million in fines to be paid by the MBTA for persistent violations of anti-idling laws (*Boston Globe* 2004). Other outcomes of this EPA decision include a detailed anti-idling compliance procedure, creating an Environmental Management System to address institutional barriers to environmental compliance, and using low-sulfur diesel fuel on all South Station Commuter Rail trains.

On the Move Coalition: Greater Boston Transportation Justice Coalition

The Anti-Idling March brought air quality and transportation issues permanently into ACE's core program set. However, the problem extended beyond the bus depots in Dudley Square. Residents were calling for changes to the packed #49 bus, constantly late rail service, and poor customer service offered by MBTA drivers. In 2000, Roxbury residents felt that many communities of color were not getting the same level of service and level of facilities. This amounted to a clear need for regional or systemic change for all transportation options throughout the MBTA service area.

In order to address these transportation issues, the T Riders Union (TRU) was formed in December 2000. TRU fought the MBTA fare increase in 2000 and especially focused on bus transfers, which were unfairly expensive. TRU took the Roxbury message to the State House, attended MBTA hearings on the fare increase, conducted letter-writing campaigns, and got media coverage, but it achieved few positive results. However, on the day of the fare increase, the community decided that enough was enough. Warren Goldstein-Gelb recollects enthusiastically that

> on a day in September 2000, a group of residents gets on a bus, the #49 bus, refusing to pay the increase in fare . . . takes that bus downtown and marches to the State House. Alright, what happens? First of all, it's the day the fare's going up so the media's going to cover the story, so they're out two hours, three hours before we even get there. They're in Dudley Square, big power, all the media there. The story is front page in the [*Boston*] *Herald* before the event takes place. OK, so we get to the State House, . . . a hundred people marching to the State House, and a delegation walks in and says, "We demand to see Governor Cellucci." We have no appointment, like "sure they're going to let us in." I think it was Gloria Fox [state representative] as part of the delegation, and his [Cellucci's] office says, "OK, we'll meet with you. We'll meet with a

> small delegation." So they met with several members of Clean Buses for Boston, including Khalida Smalls. They went in and met with the governor and said, "This fare increase is fundamentally unfair for these reasons." And the governor essentially picked up the phone and said to the MBTA, "You will negotiate with these folks. They have a point."

Alternatives Press (2000:1) reported that

> after months of organizing against the MBTA fare increase by the T Rider's Union (TRU), the MBTA's Board of Directors voted on October 12th to implement free bus-to-bus transfers by December 1st of this year. In addition, the MBTA Board voted to offer riders a weekly unlimited use, bus-subway combo pass, priced at $12.50 per week.

In this case, direct action produced results.

A more recent fare increase in 2003 motivated more action to address public transportation issues. Major changes needed to be made to the transportation system. ACE emphasized that people in low-income communities across the MBTA region should be served first because they are traditionally left out of the program. In 2000, ACE had achieved free bus-to-bus transfers and the weekly combo pass. This time around, TRU's handling of the MBTA was better, they attended MBTA board meetings and rallied on the State House steps. The sophistication of TRU's analysis was greatly increased in 2003, when it examined MBTA capital budgets. The TRU had a structured strategy in the Beat the Fare Increase campaign to address these issues, and everyone had a role to play. One of the more recent meetings with the MBTA allowed for a sharper analysis of alternative fare structures and transfer penalties. There has also been more media coverage than before, and the earlier efforts are clearly beginning to bear fruit over time. One of the important outcomes from this last campaign includes creating a T Rider Oversight Committee (TROC), of which ACE is a member, that meets monthly to discuss customer service improvements and service quality issues.

Greater Boston Environmental Justice Network

Another regional program that focuses on technical assistance to communities is the Greater Boston Environmental Justice Network (GBEJN), a network of thirty organizations that come together to work collectively on environmental justice issues. GBEJN, also created in 2000, which came out of Neighborhoods Against Urban Pollution (NAUP), is a network of community-based groups that meets quarterly about shared issues. The "EJ in the 'Hood" conference is one such quarterly meeting, and it is annually held in April. At this session, there are multiple workshops related to environmental justice, such as zoning, community master plans, organizing, and youth-led workshops. In an effort to be sensitive to community needs, ACE provides food and childcare, which are critical to creating an inclusive and hospitable environment. The "EJ in the 'Hood" conference represents the efforts of many residents in the Greater Boston area trying to find a common solution to local issues.

Northeast Environmental Justice Network

Founded in 1992, the initial meeting of the Northeast Environmental Justice Network (NEJN) was hosted by West Harlem Environmental Action (WEACT), New York's first environmental justice organization. WEACT itself was co-founded by Vernice Miller Travis, former ACE board member. Due to a number of issues, including staffing changes, NEJN has yet to fully develop a program in the way that other regional environmental justice networks, such as the Southwest Network for Environmental and Economic Justice (SNEEJ), have. A coordinator, Omar Osiris, based at WEACT, has now been hired and is beginning to develop a program for NEJN.

Safety Net

Safety Net started in reaction to both the shocking number of new developments being announced in Roxbury with little or no community input and the Roxbury master-planning process. Its aims are a "vision for a sustainable Roxbury and equitable metropolitan development." At the beginning of the master-plan process, there was an idea between Harvard and the City of Boston to create a biolab on Melnea Cass Boulevard. However, ACE talked to the local residents and found that there were serious concerns with this project and others like it that were springing up in the neighborhood. ACE began organizing local housing-development residents against the lab and held rallies to attract attention outside Roxbury. *Alternatives Press* (2002:6) reported that

> with organizing support from ACE, the Safety Net began working with residents at Grant Manor, Orchard Gardens, Madison Park and other developments to hear concerns, explain the jargon of "professional planners," and help residents voice their demands for the future of their community. They demanded that the Boston Redevelopment Authority (BRA), the city agency responsible for the Master Plan, give the residents time to have their voices heard.

In the end, Harvard walked away from the project, saying they could not handle the publicity (ACE 2004). This small victory over development interests empowered Roxbury residents, developed transferable skills, and created a renewed interest in voicing opinions for the far-reaching Roxbury Master Plan.

Safety Net is currently fighting another biolab. This is a Boston University–sponsored Level-4 biolab that will work with incurable diseases and viruses, such as Ebola and smallpox. Roxbury residents do not have the financial, political, and organizational resources necessary to combat these development issues, as other communities in metro Boston may have. This is especially true when the development proponents include wealthy universities and the Department of Homeland Security. In spite of these strong proponents, Safety Net has risen as the lead voice of opposition to the Level-4 biolab project in the past two years. Building on past project battles, Safety Net not only mobilized local Roxbury residents but also drew in regional participants. Although Boston University releases important safety information at an agonizingly slow pace, Safety Net churns out a constant stream of sophisticated economic, environmental, and social-impact analysis to raise public support. According to Michael Blanding (2004:16) in *Boston Magazine,*

> a few small voices in the South End and Lower Roxbury raised concerns about where the lab was going to be built: right in the middle of their neighborhood. And if anything went wrong, it wouldn't be just them who would be affected. It would be hundreds of thousands of people who live and work in and around this city.

ACE's Next Step: Membership

ACE's Strategic Plan (2002–2007) paves the way for the next phase of the development of the organization: membership. This is a huge and contested issue within the organization. The next step is for the organization to become composed of members who are primarily in the contiguous communities (Roxbury, Dorchester, and Mattapan) and for the programs to be originated and driven by these members. Most members, ACE hopes, will not just pay dues once a year but will be engaged in dialogue, deliberation, and ongoing conversations and ultimately will be taking active leadership in shaping the programs. They will therefore be setting the vision and strategic agenda as well as voting on the board of directors. Eventually, members will take on a greater burden of ACE's work. ACE staff can then serve as facilitators for residents acting in different programs and bring more participants into the movement.

The principal source of strength in this "new" organization will be "people power." Staff will do work to research decisions, but the membership will be the decision makers who dictate how the organization will proceed. These members will develop unique leadership roles through this work and further promote ACE's goal of community empowerment through its principal vehicles: popular education and empowerment-practice. In many ways, through membership, ACE is looking to the highest level of public participation in Arnstein's (1969) "ladder of public participation": community control.

Environmental Justice or Just Sustainability?

Having laid the groundwork through ACE's principles, popular education and empowerment-practice, through its theory of change and model of community action, and finally through its programs, I now want to look more deeply into whether ACE is still an environmental justice organization or whether its practice is now that of just sustainability or both. To determine this, I posed the question, How and why has the nature of ACE's programs and repertoires changed since it was founded? To answer it, I developed four propositions, based on some of the differences between the JSP and EJP outlined in chapter 3. I will take each proposition and examine the data.

P1: Organizations representative of the JSP show more proactivity than reactivity in their programs and repertoires toward developing sustainable communities.

That ACE is becoming more proactive in its programs and repertoires is clear from even a cursory glance at table 5.1. Bob Terrell argues from an organizational perspective that when ACE first started, its mission was to provide technical assistance to groups, whereas now ACE not only provides technical assistance to groups in both legal and organizing capacities, it is also more proactive.

For instance, Terrell says that ACE "is taking the initiative to think about specific things that we want to go after, not just responding to the Chelsea Salt Pile or the asbestos pile like putting out environmental fires but actually saying what kind of campaigns do we think make sense and actively going after them." Penn Loh makes the same point but from a community perspective. He argues that in the beginning a lot of the issues that came to ACE "were reactive and 'just say no' type of campaigns." As time progressed, however, and as ACE has developed and retained partnerships, the questions became more complex and proactive: "People have been asking the question, well what comes next? What happens when we stop the bad things from coming in the neighborhood, or when we clean up the pollution, you know, what happens to the land?"

Bill Shutkin acknowledges this reactive-proactive shift but sees it as a continuum:

> I always saw ACE as working from the defense up through that point where the folks were at the table. And that was always a core value for us, getting our constituencies to the decision-making table, be it the local zoning meeting, be it the DEP [Department of Environmental Protection] policy decision around allocation of monies for brownfields.

Shutkin also sees ACE's shift toward greater proactivity in terms of the organization's inputs to state policy development. He stakes ACE's claim for the idea of the Commonwealth's "Environmental Justice Policy" and calls it a "critical moment":

> that Massachusetts Environmental Justice Policy . . . I will, on behalf of ACE take almost full credit for that. You know, that was us. That

> was us working from '93 through the mid-90s and making shit happen. If you look in the *Boston Globe* you'll see the op-eds, the editorials from Cathy Douglas Stone about an organization called PHEED [Public Health, Environment, and Economic Development]. There's your just sustainability.

He continues by laying claim to the Commonwealth's (potential) Areas of Critical Environmental Justice Concern (ACEJC) policy: "ACEJC. That was our concept. That was my concept. I brought it up with them in 1994." The converse, however, is also true; not only is ACE becoming more proactive, it is becoming less reactive, but not entirely so. In poorer communities all over the United States (and the world), such as Roxbury, there will always be a need to react to the (many) threats that will arise because of decision makers who take the easy way out by placing LULUs in such communities. Primary among these threats at present is the Boston University Level-4 biolab proposal.

What is interesting, however, from a programmatic standpoint, is that the ACE program charged with resistance, that is, *reaction* to the proposal—the Safety Net (or more correctly, the Roxbury Safety Net)—is also the program that, on paper at least, looks like the most *proactive*, sustainable community–style program in that its aim is to "empower residents in public housing, and others in Roxbury, to develop a voice and vision for a 'Sustainable Roxbury' and equitable metropolitan development" (ACE website).

So has ACE always had, like DSNI, a vision of a more sustainable community? Loh's point, "when we clean up the pollution, you know, what happens to the land?" would seem to suggest that the question of vision, of "what's next?" resulted from communities' realization that *reaction* alone is insufficient. Goldstein-Gelb agrees:

> I think that sustainable, healthy communities has been a part of what we're doing from an early stage. . . . How do we get the community's vision of what type of redevelopment they're going to see. So that's '96, '7, '8, so that's early ACE history.

There are, however, tensions in the relationship between ACE and a more proactive, sustainable communities approach, which are less about the ends than the means. This will be more fully discussed later. Regarding the approach, Shutkin says,

> I started New Ecology for the precise reason that ACE as an organization, in terms of its competencies and capacities, didn't have the planning skills, didn't have the development skills necessary to talk seriously about whether it makes sense to turn that brownfield into housing or mixed-use commercial or a park.

James Hoyte agrees:

> What Bill Shutkin was interested in, this ecological development, sustainable development in a very explicit way, I think ACE has been reluctant to be as explicit in all that as Bill Shutkin wanted, and I think that encouraged him to want to establish New Ecology.

P2: Organizations representative of the JSP use more deliberative tools and techniques in their programs and repertoires.

With regard to community involvement in its programs, ACE has traditionally been a talking and listening organization. Khalida Smalls says that a powerful way of getting messages across is "definitely word of mouth, building relationships with people, having conversations, it's something Klare Allen and I do a lot." Building on this, Klare Allen sees every day as a teach-in: "When you talk to people, I mean when you meet people, they don't know about something, you have to quickly be able to break things down."

My starting point here is that, traditionally, EJ organizations, especially the smaller neighborhood groups, have largely been reactive in defense of their space, while just sustainability groups have been more proactive, using deliberative tools to engage people in ongoing dialogues and conversations about visions for the future. ACE as a larger, professional organization does not fully fall into this reactive-proactive dichotomy, but it is looking for more structured, continuous dialogue and conversation than it currently practices. The way to achieve this, it argues, is through its forthcoming membership drive. As Hoyte says,

> I think community involvement is one of the trickiest things we have to deal with, to tell you the truth. Quite appropriately and self-consciously, ACE as an organization is concerned—and that's why we are going over to membership—about having an ability to truly represent what our base wants to do.

Loh agrees that the move toward a more deliberative approach was not an accident:

> I wouldn't say that that wasn't part of the conscious design. In the beginning it was about establishing relationships of trust with community partners, it was about finding ways to be truly value-added to the work. But again, I think it's because of the environmental justice framework that was driving our work . . . that I think added to the more deliberative, longer-term approach.

To this extent I would argue that ACE is shifting to a more deliberative model. It may not be using many of the wide array of currently available deliberative tools, but it is using what it sees as the most appropriate vehicle in the context of the community in which it is based. Goldstein-Gelb sees greater deliberation through the membership model, working like a funnel. He explains that people are going to be involved at various levels, as with any membership organization. Some people, at the broad end of the funnel, will just say "yes, I'm a member," and that is all they will do—receive *Alternatives Press*. The theory is, however, that some people, albeit a smaller number, will move down the funnel to the middle. These will be people who have some experience with ACE, who have attended some training courses or teach-ins, who have had some dialogue with ACE, or who have joined a campaign committee or are interested in transportation, and these people are more highly engaged. Goldstein-Gelb continues:

> and then out of that group develops a smaller group that is leaders that are really maybe at the heads of these campaign committees who will then funnel into even a smaller group, which eventually is our board, so our members go up to board level.

Membership, however, has its tensions. Terrell questions what priorities and what issues the new rank-and-file members coming into ACE will be bringing with them:

> I think there is still some tension and some debate as to how and when that's going to occur, because some people, I think, take the view, "well, we've got a certain set of transportation issues that we're working on and we want to pursue that." Is that limited by or is that shaped by

> what the rank and file and TRU say, or do we pursue those issues because we know those are the right issues to go after? There's this debate about how much democracy is going to take place in the organization and who is going to make what decisions. I think there are still some tensions around that.

P3: ACE's programs are shifting from local and single focus to more regional and systemic.

Many interviewees were in basic, but not full, agreement with this proposition. Quita Sullivan says, perceptively, that "it's always been there [a regional and systemic focus]. It's a growth rather than a change." Terrell similarly sees it as part of the development of the organization:

> I think the local to regional, that's evolved. Once people started doing the work around Environmental Justice in the 'Hood, GBEJN, and the transportation work, naturally that takes you into a regional perspective. You have to, because it's in the nature of what you're up against.

Charlie Lord sees the shift, if it can be called that, as being inspired largely by Penn Loh:

> ACE was always about building power, but it was also very much about providing services to local community groups in such a way as to build that power. We also—and Penn Loh gets a lot of the credit for this—we also began to see that there was a distinction between solving problems on a problem-by-problem basis and building the power collectively to change the dynamics that created the problems in the first place.

Loh adds another dimension, that of bridging: "Yes and no. I do agree with the general . . . I think I would restate both of those, which is that I think ACE is trying to be that bridge between local and regional." Loh then uses the analogy of putting out fires. He says that rather than putting out fires one at a time, ACE is now figuring out how to achieve more systemic change, by coming up with policies and procedures that prevent fires in the first place. So, he argues, "I think there's been this focus on systemic change which has led us to be much more conscious about who we partner with and the scope of power that we're going to have to mobilize if we're going to achieve those changes." Goldstein-Gelb also sees the "bridge" and adds a caveat to my proposition. He observes that ACE's 2002–2007 Strategic Plan does talk about systemic change on a regional basis but that it also (re)roots the organization back in membership and having local control. He also sees two tensions or challenges. The first is the challenge of getting local people active on regional issues:

> They say "the 23 bus is overcrowded and really busy, and I want something addressed there," and so clearly we want to work, as our organizing strategy, on the 23 bus. And we also want to have folks see the larger picture. So there is a bit of . . . inherently I think there's a tension there and one of the challenges of organizing on both a local and regional basis. That's going to be a challenge for us.

The second challenge Goldstein-Gelb sees is

> getting the regional folks to respect and focus on the grassroots constituents, because there's also—this is a real problem for us—there's always the risk of plunging ahead into the systemic stuff and saying, "well, we don't have the residents with us right now because they're focused on their local issues but we know what would be good for them." That is a tension that comes up over and over again.

P4: ACE is more likely to build coalitions with (other) organizations representative of the JSP than those representing an environmental sustainability orientation.

Gould et al. (2004) have shown through their work on "Blue-Green" or "Seattle" coalitions that environmental justice and environmental health organizations are more likely to form alliances with labor than are habitat-focused or environmental sustainability groups. My proposition is that ACE's coalitions are more likely to be with like-minded groups such as the Bowdoin Street Health Center, DSNI, the Washington Street Corridor Coalition, the Coalition to Protect Chinatown, the Four Corners Action Coalition, City Life/Vida Urbana, Bikes Not Bombs, and the Boston Tenant Coalition than with habitat-focused environmental sustainability groups such as the Massachusetts Audubon Society, the Sierra Club, or the Conservation Law Foundation.

In ACE's case, this turned out to be precisely the situation with regard to their longer-term coalitions (Gould et al. 2004:90), although there was one interesting exception: the Boston Group of the Sierra Club. There were concessions that "short-term marriages of convenience" (Gould et al. 2004:90) could potentially be with less like-minded groups, but these tended to focus on specific, time-limited issues.

Loh sees coalitions as an acceptance of ACE's principles, a commitment to a longer-term project, and to movement building:

> The way we do coalition work is pretty deep. We tend not to do stuff that's, like, coalition means signing on to something and then it's over in a year. So I'd say the way we approach coalition building doesn't lend itself to those short-term marriages, although I think that we have been fairly successful in attracting the alliance of folks who do get it, and I think that's part of spreading the EJ equity message.

Interestingly, the research findings of Gould et al. (2004) regarding labor organizations and their closeness to EJ and environmental health organizations, as opposed to environmental sustainability or NEP groups, has a direct bearing on ACE's coalition work today. Sugerman-Brozan says,

> We are increasingly looking for ways to connect with labor groups, and in Dorchester, the Dorchester-Roxbury Labor Committee is going with this issue of not enough jobs for community folks on its whole T station renovation, so we're connecting with them . . . the idea is that we're connecting with them to bring them more into On the Move and thereby a longer-term coalition.

Unfortunately, this connection was very recent and, at the time of my case study, was unshaped. Lord, like Loh, sees the movement-building aspect of coalitions and adds another vital element, namely, funding. He notes that ACE's first coalitions, especially the Neighborhoods Against Urban Pollution project, were a response to an emerging understanding that there were similar problems in a number of Boston neighborhoods. In tackling these problems, ACE needed a variety of supports and help in policy-related disciplines, so it made links with toxics groups, with the Tellus Institute, which had skills in economic analysis, and with groups like DSNI and the Bowdoin Street Health Center that had a local organizing interest and similar challenges. It was therefore apparent that ACE said, in effect, "let's try to work together to understand these challenges and also to develop methodologies to address them."

Lord continues:

> There was also a pragmatic component, funding. Rather than compete for an emerging base of funding for EJ in Boston and nationally, we felt that, and I think ACE still does feel, that it was more efficient to find like-minded organizations and work together to find the resources we need to tackle these underlying issues.

Goldstein-Gelb adds yet another dimension in that "it's not necessarily that there are more coalitions now, but that they're longer-lasting coalitions. Those [earlier coalitions] were more temporary issue-specific coalitions." In effect, what he is saying, and this backs up Loh's point, is that they are focusing more on "deep" coalition building today.

However, this does not preclude coalitions "lite." Sugerman-Brozan talks about short-term coalitions and notes that they can be with non-progressive groups:

> Short-term coalitions for us, I think, tend to be single-issue coalitions where, for instance, with the [Beat the] Fare Increase . . . we were trying to build connections to groups like Mass Taxpayers Foundation, who were trying to call on the T to have better fiscal responsibility. They are not a progressive group. They are not a group that has any long-term vision that is similar to ours, but for that particular issue it was important to bond with them to address it.

In response to a probe about the possibility of working with more environmental sustainability or habitat-based organizations, there was a definite ACE line. Shutkin said that such groups are "almost irrelevant. . . . Unless something's changed, the real traditional natural resource, wildlife organizations, of which there are a few in Boston, have never really played much of a part in ACE's life." ACE takes this line because, as Lord said earlier, "the environmental movement . . . was missing an element of social justice, which meant a lot to me." However, there is one notable exception to the ACE line: the Boston Group of the Sierra Club, whose national organization scores a JSI of 2. Terrell notes,

> We're fortunate. The Boston Group of the Sierra Club is very progressive, and we have a good working relationship with them. But that has come through my organization the Washington Street Corridor Coalition

> [WSCC]. They have been a part of our coalition for years. So when we got involved with Clean Buses for Boston and On the Move and all that stuff, we kind of brought them along with us as part of our coalition.

The history of the WSCC's relationship with the Boston Group of the Sierra Club is interesting and bears mention here since it ties in with some of the formative arguments in the development in the early 1990s of the environmental justice movement and its relationship to the traditional environmental sustainability movement. Terrell recalls that the Sierra Club in the early 1990s was going through some real changes at a national level:

> Some of the local people in the Boston chapter, who had been transit activists, came to us and said, "Hey, we agree with what you're doing around Washington Street light rail and some other transit issues." They were very much involved in the North-South rail link, and so they came into the WSCC very early in our history, and they've been good allies. I mean they've been extremely principled in the way that they work with us.

Terrell then links this local allegiance to the wider national debates of the early 1990s, namely, the "Letter to the Big Ten" and the 1991 People of Color Environmental Leadership Summit, and finally to the relationship between sustainable development and environmental justice. I quote him at length here because I think what he says is pivotal and catches the tension and opportunity arising at that time:

> The "Letter to the Big Ten" had a tremendous impact on the [People of Color Environmental Leadership] Summit because people came ready to do serious political battle at that conference. I mean they were just itching to get at the Sierra Club, National Audubon, National Wildlife Federation, Defense Fund, and all those people. But at the local level, that was just about the time that the Sierra Club folks approached the WSCC and said, "You know, we're on the same page on this stuff. Let's talk; let's work together." I think John Deakin and John Lewis came into the WSCC around 1989 and, 1989 to 1990 thereabouts, and when the [People of Color Environmental Leadership] conference took place in Washington, D.C., John Deakin arranged for me to come to the

> Sierra Club Boston chapter board of directors and do a presentation on our perspective on EJ, and our perspective that came out of the '91 [People of Color Environmental Leadership] conference, and my presentation was very theoretical. It was on a very theoretical and historical level, but I did that deliberately to let them know that our conceptualization of sustainable development and environmental justice came out of years of theoretical and practical work in our communities and that this conception didn't just drop out of the sky yesterday. It wasn't just because people wanted a chemical plant pushed out of their neighborhood; it was a lot deeper. There was a lot more thought. It wasn't just "get something out of my back yard."

Sugerman-Brozan, herself a former worker in the traditional environmental sustainability movement, identifies a key point in this local "brown-green" alliance, namely, the role of individual relationships, which Terrell alluded to in mentioning the Boston Group of the Sierra Club's John Deakin and John Lewis but did not expand on. In the following quote, the person that she mentions at the Greater Boston Group of the Sierra Club, Massachusetts chapter, is Jeremy Marin, Sprawl and Transportation Organizer:

> I think it really varies with who's on staff at those groups to a point. I mean Sierra Club have definitely been part of different coalitions because they support light rail on Washington Street, but right now there's someone who's a staff person in Boston who really gets it, who really cares about EJ. And so you see the Sierra Club popping up a lot more, supporting EJ positions a lot more than we did in the past.[5]

At the same time, Sugerman-Brozan is aware of the problems:

> Even though the [Boston Group of the] Sierra Club has been very strong in supporting a lot of the work that we've done on transportation, they can't sign on to our platform, so they're not officially members of On the Move at all . . . but there are certainly places where they have been helpful, and there are places where we have run into problems.

Indeed, on the website of the Greater Boston Group of the Sierra Club, there is mention of their transit activism:

> MassTransit
> We have been working on environmental and legal issues concerning the extension of the Old Colony Line, Greenbush Line, Arborway Line and the replacement of the light rail along Washington Street to Dudley Station. (http://www.sierraclubmass.org/groups/gbg/gbg.html)

However, there is no mention of the On the Move Coalition, backing up Sugerman-Brozan's point. Similarly, an op-ed in the *Boston Herald,* jointly penned by Jeremy Marin and Bob Terrell, dated February 16, 2003, and entitled "Light Rail Makes More Sense for Washington St. Corridor," makes no mention of it.

This link between ACE and the Boston Group of the Sierra Club is a hopeful sign that, despite problems that need to be addressed at the local and individual level, more so than at the national level, longer-term coalitions (Gould et al. 2004) or what might equally be called "co-operative endeavors" (Schlosberg 1999) are possible across race, class, and paradigm lines. This is a theme to which I shall return in chapter 6.

The picture is altogether less rosy with regard to ACE's dealings with another group, which, like the Boston Group of the Sierra Club, is firmly within the NEP. The most powerful group on the Boston environmental scene, with a JSI of 1, the Conservation Law Foundation (CLF) is

> the oldest and largest regional environmental advocacy organization in the United States. We're based in New England, where our attorneys, scientists, economists, and policy experts work on the most significant threats to the natural environment of the region and its residents. We've been around since 1966, and maintain advocacy centers in Boston, Massachusetts; Concord, New Hampshire; Montpelier, Vermont; Providence, Rhode Island; and Rockland, Maine. (http://www.clf.org/aboutclf/index.htm)

CLF's focus, if we are to read its mission literally, is first the "natural environment of the region" and then "its residents." Its program areas are agriculture, clean air and climate change, communities, marine resources, natural resources, and "other areas of interest" such as the "transportation for livable communities network."

ACE's problem with CLF is basically ideological, related to their different models of community action and theories of change and to ACE's deeply held principles. As we have seen, ACE has a bottom-up model of community action, whereas CLF's model is more top down, more expert led. The difference is summed up by Sugerman-Brozan:

> Their tactic is about gaining access to the decision-making structure and then working from the inside. Our tactic is about building power from the ground up. So when you have those different approaches to building power, you have a very different way of doing things.

This difference surfaced initially through ACE's experience on the Clean Buses for Boston coalition. Loh says,

> CLF was an original founder of Clean Buses for Boston, but we asked them to leave because the way they worked didn't match. . . . They were willing to drive their own agenda without consulting anyone else, so for us, we didn't feel comfortable keeping them in that close circle.

With regard to their theories of change, CLF and ACE are poles apart. Loh calls the CLF approach "environmental imperialism" and argues that ACE

> came into this work . . . through an EJ framework which has an explicit analysis of race and class and how environmental issues play out through those lenses, and I think CLF probably came to the work without that consensus. And I'm not just saying CLF; there's a lot of other organizations in the environmental movement that are the same thing.

Sugerman-Brozan really pins down the problem, the conflict between the reformative politics of CLF and the transformative politics of ACE. She says, "I think they don't see the whole system as unjust and needs fixing, I think that they see that you can change it and get adequate results." Loh's point about the environmental imperialism of not just CLF but "a lot of other organizations in the environmental movement" and Sugerman-Brozan's point regarding reformative versus transformative politics go to the heart of the problem.

CLF, like the majority of other organizations in the traditional movement, is an environmental sustainability organization working within the heart of the NEP. Although not entirely habitat focused, it is content to tweak existing policies while maintaining the status quo. ACE, on the other hand, is a transformative organization, working within the

EJP/JSP in disinvested communities toward redistribution, toward paradigm shift. CLF does work on "transportation for livable communities" such as

- making neighborhoods safer, quieter and more child-friendly through traffic calming;
- bringing new life to old commercial areas with public transit improvements; and
- improving local mobility by retrofitting streets for bicycling and walking. (http://www.clf.org/transportation/index.htm)

Their focus, however, is not, like ACE's, explicitly on equity and justice, although this may be implicit. Nor was it revealed to me, in the way that Terrell had revealed to me regarding the Boston Group of the Sierra Club, that there were possibilities of rapprochement through personal relationships. Interestingly, however, ACE's former executive director, Veronica Eady, was appointed in summer 2004 to the CLF board of trustees and board of directors.

Conclusions

The question I now need to ask is as follows: is ACE still an environmental justice organization following the EJP, or is its practice now that of just sustainability guided by the emergent JSP? Based on my study, and especially the responses I received to questions based on my four propositions, I could argue that ACE is an organization that now operates fully within the JSP. By the same token, I could argue, as Bill Shutkin, Penn Loh, Warren Goldstein-Gelb, Bob Terrell, and James Hoyte did, that ACE has *always* been about just sustainability, but they chose not to call it that.

Perhaps, as Quita Sullivan perceptively argued, ACE has not so much changed as grown. In this sense, the seeds of sustainability—and more specifically, just sustainability—were always there, but the discourse of environmental justice, full of the rhetoric of civil rights, was the more powerful discourse in a community such as Roxbury, because it was "home grown" in similar communities facing similar issues around the United States. This line of argument fits well with one of the mantras of the EJ movement: "We speak for ourselves." Put another way, why adopt someone else's discourse when you have your own? There is compelling evidence that ACE sees the issue this way.

First, it became apparent during my research that there is a strong reticence among ACE's staff and board members to engage in the (white, middle-class-mediated) sustainability discourse, even though in practice, a just form of local sustainable development is clearly what ACE, and its local partner DSNI, is and has been doing.

Hoyte and Terrell see this very clearly. Hoyte argues,

> I think that the sustainability movement, President's Council [on Sustainable Development], and all the rest of them had mainstream involvement, grounding. Those in the environmental justice movement were suspicious even though what environmental justice folks were doing was very akin to or had strong elements of sustainability, but to identify in that way was problematic for them.

Terrell is more specific:

> Some people have a conception of sustainability or sometimes they call it environmentalism, sometimes they even call it smart growth—they call it all kinds of different things—but because the conversation started in a white, middle-class environment, they define it in white, middle-class terms.

Second, Loh makes an argument about what he sees as the peak of the sustainability agenda in the United States in the mid-1990s, around the time of maximum activity at the President's Council on Sustainable Development. Essentially, Loh is saying, like Hoyte and Terrell, that there was a struggle over the definition and ownership of the sustainability agenda, but he adds that the EJ movement needed to "stake its claim":

> they were saying that there's a sustainable development dialogue going on, which at that time was much more prevalent, especially at the federal level. There were all these people . . . what they were trying to do was say "what's our own take on what sustainability is?" you know, "we need to come up with our own definition and claim our space on this." So there was a whole dialogue and result and report-back on what that meant to communities

Loh then goes further, arguing that the origins of just sustainability lie in the environmental justice movement:

> the roots of just sustainability, I'm not going to say that it was exclusively owned by EJ, but I think people in the EJ movement, it was conscious, there were conscious efforts to make those connections from fairly early on, as far as I know.

In support of this assertion, Loh showed me a copy of *Sustainability and Justice: A Message to the President's Council on Sustainable Development,* produced by the Urban Habitat Program in April 1995. In its introduction, Carl Anthony, Hannah Creighton, and Loh argue that

> the groups who make up the growing environmental justice movement . . . are indeed leading the way towards sustainability in the U.S. . . . sustainable development requires a fundamental reorganization not only of the way we produce and consume, but the way that we govern and regulate these basic activities. (Urban Habitat Program 1995:1)

A final point about how ACE views itself involves my argument throughout this book that justice and sustainability are inseparable. In response to the question "is ACE an EJ organization, is it a sustainable development organization?" Terrell responds,

> I think ACE is both because . . . we come out of the context of the African American community and other communities of color. Most of us doing this kind of work see sustainability and environmental justice as two sides of one coin, but you really can't have one without the other. Now, we understand why other people think differently. To some people it's justice—"get it out of my neighborhood"—that's it, or sustainability people, who have wonderful ideas about solar energy.

Terrell then continues with a very specific line of reasoning about the links between justice and sustainability, using affordable housing as an example. Again I quote him at length, as his message is so germane to this book. He sees ACE's future as a lead organization in this justice-sustainability linkage:

> I think one of the challenges to ACE over the next ten years is that, as we take up those campaigns and as we do things out in the neighborhood and we think about proactive strategies for the neighborhood, sustainability is going to have to be a huge part of that conversation. For example, we've already said we're all in favor of affordable housing. ACE participated in the Roxbury Master Plan. We talked to other people in the neighborhood about where the housing should be built, how affordable it should be. Now we've got to ask how energy efficient is that housing because 30 percent of your costs of housing is wrapped up in energy. So, how do we go about promoting housing that is energy efficient? Do we start to promote more solar energy technology as being a part of the housing that we build in the future? And this is a big discussion we have to have with contractors, real estate developers, and community development corporations. We need to integrate that thinking into our housing production. ACE is perfectly situated to take the lead in having that kind of conversation, and that's a very radical notion. Most of the affordable housing people are not thinking along those lines.

ACE could be called an environmental justice organization. It could also be called a just sustainability organization. What is certain though is that it is in the vanguard of a movement for just sustainability that is looking to integrate justice and sustainability in a practical, grassroots way by building power in Roxbury, metro Boston, and the wider New England region.

6

From Confrontation to Implementation

Some Concluding Thoughts

In this final chapter, I present five concluding thoughts, out of the many I could have chosen, which I hope will help fuel the debate that will begin to move us from the tradition of confrontation between the EJP and the NEP toward greater understanding and hopefully strategies for the implementation of robust, joined-up policies for just sustainability at the local, regional, national, and international levels.

First, I develop thoughts on the debate about whether the local or national level is a better place to make coalitions or engage in the cooperative endeavors that I believe are leading the way toward more just and sustainable futures. Second, and related to my first point, I argue that the best chance for more cooperative endeavors, and ultimately movement fusion between the EJP and NEP, will come from environmental justice groups working with just sustainability groups (JSI = 3), as opposed to those of an environmental sustainability orientation, but that local issues also come into play that can change this. Third, I examine the problem of unfulfilled policy rhetoric regarding the implementation of just sustainability projects and programs. Fourth, in pursuit of just sustainability, I conclude that there is an emerging set of tools such as environmental space that show exceptional promise for the JSP and the EJP in the United States and worldwide. Finally, I revisit my question of whether ACE and others like it are environmental justice or just sustainability organizations.

Local or National?

Are coalitions best made at the local or national level? Common sense would tell us that the answer is contingent on several factors, such as the issues being pursued, the targets of the coalition, the ideological and interpretive positions of the organizations, the coalitional time span, and many other context-specific factors. For example, "The Johannesburg Summit 2002: A Call for Action" was led by Redefining Progress, but that organization was joined by major organizations such as the Natural Resources Defense Council, Friends of the Earth, the Earth Policy Institute, the Worldwatch Institute, the World Wildlife Fund, the Environmental Law Institute, the Rainforest Action Network, the Nature Conservancy, Greenpeace USA, the Sierra Club, and the Woods Hole Research Center. This was a national coalition focusing on what the organizations perceived to be a national (lack of) policy in the run-up to the WSSD in 2002.

In the words of Gould et al. (2004:96), it was a "short-term marriage of convenience," convened to deliver a specific sustainability policy message, aimed at the Bush administration's vacillation on the issue. At this level, the issues, targets, time span, and sheer urgency of the "Call for Action" overrode any ideological or interpretive differences among the individual organizations within the coalition. Yet there clearly are discursive, ideological, and interpretive differences among these organizations, because Redefining Progress is an organization with a JSI of 3 in coalition with a mixture of other organizations with varied JSIs including one with a JSI of 0 (the Nature Conservancy).

Another issue, that of light rail on the Washington Street Corridor in Roxbury, with a different target, the MBTA, elicited a different, more local, and longer-term coalition between the WSCC and the Boston Group of the Sierra Club. While clearly there were individual relationships between the staffs and especially between the executive directors of the national organizations in the pre-WSSD "Call for Action," ACE's Jodi Sugerman-Brozan identified individual staff relationships as being key to the endurance of the local coalition of ACE, WSCC, and the Boston Group of the Sierra Club, even though ideological and interpretive differences precluded the Boston Group's "official" membership in the On the Move platform and coalition.

It seems therefore that coalitions can, in certain circumstances, override discursive, ideological, and interpretive differences, especially but not exclusively in the shorter term. However, as ACE's Penn Loh mentioned, ACE's approach to the "longer-term coalitions" of Gould et al. (2004:96) is "pretty deep," by which he means that the ideological and interpretive heart or focus of the coalition must be on equity and justice. This is why Loh decided that ACE would not be part of the Massachusetts Smart Growth Alliance:

> We chose not to become part of the Smart Growth Alliance because we didn't feel that the justice issues were front and center enough to make it worth it. We don't want to spend time, even if there are some opportunities to be had there, fighting within the partnership itself to ensure that the approach is right.

Put another way, "equitable development is about who receives the benefits and burdens of development as well as where the development happens. . . . Too often, smart growth focuses only on the where" (Loh, quoted in Marsh 2003:1).

Justice as the Focus?

Following from Penn Loh's point about coalitions, my argument is that the best chance for "longer-term coalitions," cooperative endeavors, and ultimately movement fusion between the environmental justice and sustainability movements will come from environmental justice groups working with just sustainability groups (JSI = 3) as opposed to those of an environmental sustainability orientation (JSI = 0–1). This was certainly the case with ACE and its longer-term coalition partners such as DSNI, the Coalition to Protect Chinatown, and City Life/Vida Urbana. It also explains ACE's problem with CLF's approach, described by Loh as "environmental imperialism." However, as I note earlier, this EJP-JSP, as opposed to EJP-NEP, alignment may not always be the case when all partners perceive that action is urgently required, as in the case of the pre-WSSD "Call for Action," or when the combination of a local issue and enduring individual relationships, such as that of ACE, WSCC, and the Boston Chapter of the Sierra Club, makes a coalition a sensible option despite ideological and interpretive differences that preclude "official" coalition membership.

In relation to environmentalist-labor coalitions, which I have argued raise similar problems to NEP-EJP coalitions, Gould et al. (2004:109) note that

> more conservative groups like the National Wildlife Federation, the World Wildlife Fund, and the Nature Conservancy appear to have all

> been uninterested in such coalitions. On the other end of this spectrum, some have already said yes, including the Sierra Club and Friends of the Earth.

This backs up my arguments, and supports the rankings given to organizations in my JSI, in that the National Wildlife Federation, the World Wildlife Fund, and the Nature Conservancy have JSIs of 0, 1, and 0, respectively—that is, they are firmly within the NEP—whereas the Sierra Club and Friends of the Earth both have JSIs of 2, which, while not enough to qualify them as fully adhering to the JSP, does highlight their more progressive nature.

I have argued throughout this book that the issue of equity and justice unites the EJP and JSP, although groups representative of the latter may not have experienced injustice in the personal and visceral way that many neighborhood-based EJ groups have. In this regard, it is no coincidence that the report "African Americans and Climate Change: An Unequal Burden" (CBCF 2004) was commissioned by the Center for Policy Analysis and Research, the policy arm of the Congressional Black Caucus Foundation, Inc. (CBCF), but was conducted by Redefining Progress, a JSI 3 organization.

While CBCF is not an environmental justice group per se, its concern for African Americans and climate change is an environmental justice issue. Therefore, the choice of Redefining Progress as a partner makes a lot of sense. It has a track record of credible research, and its mission states, "Redefining Progress is a non-partisan public policy institute focused on the intersection between economics, social equity, and the environment." Its executive director since 2001, Michael Gelobter, is African American and a former professor at Rutgers University, with both a research interest in environmental justice and a respected position within the environmental justice movement. As I showed in chapter 4, there are not many national environmental and sustainability organizations that take equity and justice as seriously. CBCF's report, which forecasts a difference in both the timing and magnitude of the impact of climate change on people of various socioeconomic and racial groups, argues that "where U.S. energy policy is concerned, African Americans are proverbial canaries in the mineshaft" (CBCF 2004:5).

Interestingly, in relation to the CBCF report and to my comments in chapter 1 about the lack of joined-up policymaking in Massachusetts, the state's "Climate Protection Plan" (Commonwealth of Massachusetts

2004), though published before the CBCF report, makes no mention of environmental justice or the state's policy. The fifty-four-page document does, however, mention the word *sustainable* sixty-nine times, *sustainability* nineteen times, and *sustainable development* twenty-two times, an average of just over two mentions per page. The policy was developed by the OCD, which, ironically, is responsible for "the integration of energy, environmental, housing, and transportation policies, programs, and regulations" (Commonwealth of Massachusetts 2004).

For those who might argue that the EJP/JSP focus on equity and justice is purely anthropocentric, I would add that Low and Gleeson's (1998:157) concept of *ecological justice,* which they argue is "justice to non-human nature," applies. The Principles of Environmental Justice have a very strong bioethical strand, although as I mentioned earlier, the anthropocentric tradition has become dominant for good and understandable reasons.

Just Sustainability: Reality or Rhetoric?

While I have demonstrated clear evidence of the emergence of the JSP in some national and international organizations and within the practice of some progressive national governments, there is far more action in programs, projects, and organizations at the local level. There is also, however, a lot of rhetoric. Warner's (2002) research into just sustainability highlighted the progressive nature of San Francisco's sustainability policy in that it reached what he considered the highest level of commitment, "implementation," that is, the integration of environmental justice issues into its sustainability plan. This sounds impressive, and it looks so on the city's website. However, Portney (2003) saw it very differently. Recognizing the tensions in the relationship between the City of San Francisco and the nonprofit Sustainable San Francisco, he concluded that despite the latter convincing the city to adopt the sustainability plan as policy, he could find little hard evidence of its implementation.

Again, it is in the smaller, local projects, such as my vignettes illustrating a wide range of just sustainability concerns and the work of ACE in Roxbury, which are leading the way toward showing how just sustainability can be and is being implemented.

Policy Promise: Environmental Space

Like the Precautionary Principle and clean production, environmental space is an area of exceptional policy promise for the JSP and the EJP, in the United States and worldwide. McLaren (2003), now the director of Friends of the Earth Scotland, has argued that environmental space underpins just sustainability and shows how sustainable development policies can incorporate equity and justice issues in an explicit, quantifiable manner, rather than the implicit manner that is more common in the NEP. Crucially, therefore, environmental space quantifies and helps operationalize sustainability while simultaneously highlighting the role of equity and justice.

There are problems with the concept, however. The very name hints that its main concern is based around "the environment" and environmental resources, when, as I have argued, sustainability is about more than only "green" issues. One could say that its "social" aspect is that it is about the equitable and just distribution of these resources. As Buhrs (2004:433) argues, "although the ecological and resource dimensions are essential for any operationalization of sustainability, there is also a social dimension that is much harder to define, quantify and operationalize." He continues by arguing that the indices I mentioned earlier with a social or well-being focus, such as the Human Development Index, the Genuine Progress Indicator, and the Index of Sustainable Economic Welfare, while useful, are difficult to translate "into more or less specific goals or targets, alongside those for resources and sinks" (433). This is a challenge, but it is not an insurmountable one. A research agenda could be built around a composite "sustainability" or "community space" index that attempts to define a floor and ceiling to not only environmental resources and sinks, as in environmental space, but to social and economic entitlements too.

There are increasing signs from local, national, and international experience that environmental space analysis, notions of "fair shares" and of per capita environmental rights, are gaining ground. Sachs (1995) argues for greater links between human rights and environmental rights, and, as Bosselman (2001) points out, by the end of the last century just under sixty countries mentioned environmental rights in their constitutions. However, as Buhrs (2004) perceptively argues, environmental rights generally deal only with *minima*, basic human entitlements or the "floor," of environmental space. In order for rights to have the teeth in environmental space terms to control resource consumption, they will need to prescribe *maxima,* or the "ceiling," above which consumption is unsustainable. Doing this in different locations and countries will bring us back to the wider policy discussions, touched on in chapters 2 and 3, regarding how we measure progress and success, need, sufficiency, and efficiency.

Buhrs (2004) sees three main avenues for the institutionalization and implementation of environmental space: through the legal-constitutional framework (as rights and obligations); through the economic system (as environmental property rights) and through green planning (as objectives and targets within national environmental plans). The U.S. reader, more than one in, say, the Netherlands or Denmark, may be feeling that all hope for institutionalizing environmental space in the U.S. political system is lost at this stage, given the realities of both current Republican and Democrat thinking.

However, it may be that the vehicle most appropriate for raising (or starting?) the debate in the United States is based in our "environmental justice infrastructure" of an executive order, NEJAC, and the growing number of state policies shown by the American Bar Association's (2004) survey, such as that of Massachusetts, that are beginning to identify environmental justice populations based on key criteria but that are, in terms of specific redistributive targets, light or nonexistent. No other country in the world has this infrastructure nor such an advanced EJP and movement.

There are two priorities on which the environmental justice and emerging just sustainability movements should consider cooperating. First, at the national level, the words "fair treatment," used in the EPA definition of environmental justice, could be supplemented with the concept of "fair share" and some supporting text, which would open up a whole new horizon for debate around environmental, sustainability, or community space targets in these and other populations. Second, at the state level, an environmental, sustainability, or community space approach to clearly defined environmental justice populations would force the joining up of policy debates. With its clear reference to procedural, substantive, and distributive justice, the Massachusetts policy is ripe for molding along environmental space lines. Would or even could the Massachusetts Climate Protection Plan have ignored the substantial environmental justice implications shown by CBCF (2004) if the state had taken an environmental space approach?

Environmental Justice and Just Sustainability?

As I mentioned in the book's introduction, the JSP is not rigid, single, and universal. It links the EJP and NEP and is best visualized as flexible and contingent, with overlapping discourses that come from recognition of the validity of a variety of issues, problems, and framings. In this sense, the JSP acts like a bridge between the EJP and the NEP.

I have developed five differences between environmental justice and just sustainability: the JSP has a central premise of developing sustainable communities; the JSP has a wider range of progressive, proactive, policy-based solutions and policy tools; the JSP is calling for and has developed a coherent "new economics"; the JSP has much more of a local-global linkage; the JSP is more proactive and visionary than the typically reactive EJP. These broad differences should be seen in the context of my points about the malleability of the JSP. I am not saying that the EJP has none of these features but that the JSP has emphasized or developed them to a greater extent. The point I wish to leave the reader with is the complementarities of the EJP and JSP. Perhaps, rather than asking the question "environmental justice *or* just sustainability?" I should follow Bob Terrell's assessment that ACE is both an environmental justice *and* just sustainability organization. Indeed, I believe that ACE's success in both arenas is in large part due to the division within the organization that I mentioned in chapter 5.

Perhaps ACE is, in Lichterman's (1995) terms, a "multicultural alliance," not between two organizations but rolled up into one organization. And this was, according ACE co-founder Bill Shutkin, no accident; it fits the ACE theory of change. Terrell shares this view and adds,

> It's what some people refer to as "prefigurative politics." The way that you build your organization, the way you build your movement reflects not just your values but the kind of society you ultimately want to live in. I think there's a lot of that in ACE.

The logic of ACE as a multicultural alliance is compelling for reasons I explained in my case study, but there are two additional aspects to this. First, at the organizational level, the organizers out in the community are largely a reflection of the demographics of the Roxbury community: African American, Latino, and white. They speak the language of the local community and practice a more communitarian form of movement community building, where, as Khalida Smalls said, "word of mouth, building relationships with people, having conversations, it's something Klare and I do a lot." Klare Allen, building on this, said that in her community work the messenger is more important than the message.[1] On the streets of Roxbury, it is who you are, not so much what you have to say. This, in Lichterman's (1995) terms, is the exact opposite of the more personalized form of movement community, where it is the message (a green philosophy, personal empowerment) that is more important than the messenger.

The office-based staff of ACE is less representative of the Roxbury community and more professionally qualified, with bachelor's or master's degrees, than are the organizers. While clearly deeply sympathetic and ideologically committed to the communitarian form of community building (otherwise they would not work there), their major contribution to the organization is through their personalized form of movement community, based on their own empowered status, through which they leverage financial, media, and legal power and influence. These resources are brought to the organization and then "distributed," if this is the right word, through the organizers in the community, such that the community (and wider metro region) gains tangible benefits like, for example, funding for Air Beat pollution-monitoring equipment in Dudley Square; stricter regulations on dumpster storage lots, junkyards, and recycling facilities to stop health code violations; free bus-to-bus transfers; 350 compressed-natural-gas buses; or more "face time" for community members with local and state officials.

I do not, however, want the reader to think that it is only the office staffers that can accomplish these things. They may be able to negotiate with gatekeepers, but several of my interviewees said that ACE's credibility with funding agencies and other power brokers was based on both its keen eye for financial, legal, media, and policy windows and opportunities and its solid community-based organizing. For instance, Warren Goldstein-Gelb says,

> On the one hand we have attorneys, who are willing to work inside the system and negotiate. That negotiation, we would argue I think, is only as strong as the perception and the reality that there's a movement of people outside, banging on the door.

Similarly, James Hoyte acknowledges the leveraging power of the office-based staff and adds that the organization needs "the presumed support of a community base, and that's why it's so important that it doesn't look like the emperor has no clothes."

Again, as I said in my case study, ACE works as an organization because of the mutual respect within the organization for different roles and relationships across the divide. My feeling is that ACE's success, its credibility as an organization, is in large part because it is, at the organizational level, a multicultural alliance, able to skillfully use the overlapping discourses of both the EJP and the JSP and to translate these into practical, tangible strategies for change at the local level.

At the supra-organizational level, ACE has generated a wave of goodwill among many supporters and well-wishers, including Boston power brokers who assist ACE in the resource-gathering process. As Lisa Goodheart argues,

> I think ACE is unusual in terms of the spectrum of people that are committed to it, and it is a community-driven organization. But you have a lot of people like me that are lawyers at downtown law firms, other service providers. You have professors at prestigious institutions. You have a lot of people in traditionally powerful and prestigious positions who are also dedicating a lot of resources. I think that is different than is the case with a lot of other organizations. That gives ACE a different depth of resources.

My argument throughout this book has been that the vehicle best suited to rapprochement between the EJP and the NEP, to cooperative ventures, to multicultural alliances, and ultimately to movement fusion is the emerging discursive frame and redistributive paradigm of just sustainability, with its trenchant analysis of race and class, justice and equity, together with a solid base in limits-based environmental protection (however unpalatable "limits" may be in the United States).

ACE comes very close to achieving this synthesis, this alliance *within* one organization, and other not-for-profits wishing to do so might want to look to ACE for ideas. In terms of rapprochement between EJP and NEP organizations, ACE's Khalida Smalls offers an olive branch, a supportive and optimistic note regarding the Sierra Club, albeit a progressive NEP member:

What's the one we work with now, the Sierra Club? Are they greenies? We have to recognize that they are going through different transitions too. I think they're really beginning to broaden their view about what it is that's important, that they need to be paying attention to, and environmental justice is definitely one of those things. Just where it goes over the next few years . . . I can't see it, but I know it's bound to be the next level of evolution.

Appendix
Principles of Environmental Justice

WE, THE PEOPLE OF COLOR, gathered together at this multinational People of Color Environmental Leadership Summit to begin to build a national and international movement of all peoples of color to fight the destruction and taking of our lands and communities, do hereby re-establish our spiritual interdependence to the sacredness of our Mother Earth; to respect and celebrate each of our cultures, languages, and beliefs about the natural world and our roles in healing ourselves; to insure environmental justice; to promote economic alternatives which would contribute to the development of environmentally safe livelihoods; and to secure our political, economic, and cultural liberation that has been denied for over 500 years of colonization and oppression, resulting in the poisoning of our communities and land and the genocide of our peoples, do affirm and adopt these Principles of Environmental Justice:

1. Environmental justice affirms the sacredness of Mother Earth, ecological unity and the interdependence of all species, and the right to be free from ecological destruction.

2. Environmental justice demands that public policy be based on mutual respect and justice for all peoples, free from any form of discrimination or bias.

3. Environmental justice mandates the right to ethical, balanced, and responsible uses of land and renewable resources in the interest of a sustainable planet for humans and other living things.

4. Environmental justice calls for universal protection from nuclear testing, extraction, production, and disposal of toxic/hazardous wastes and

poisons and nuclear testing that threaten the fundamental right to clean air, land, water, and food.

5. Environmental justice affirms the fundamental right to political, economic, cultural, and environmental self-determination of all peoples.

6. Environmental justice demands the cessation of the production of all toxins, hazardous wastes, and radioactive materials, and that all past and current producers be held strictly accountable to the people for detoxification and containment at the point of production.

7. Environmental justice demands the right to participate as equal partners at every level of decision making including needs assessment, planning, implementation, enforcement, and evaluation.

8. Environmental justice affirms the right of all workers to a safe and healthy work environment, without being forced to choose between an unsafe livelihood and unemployment. It also affirms the right of those who work at home to be free from environmental hazards.

9. Environmental justice protects the right of all victims of environmental injustice to receive full compensation and reparations for damages as well as quality health care.

10. Environmental justice considers governmental acts of environmental injustice a violation of international law, the Universal Declaration on Human Rights, and the United Nations Convention on Genocide.

11. Environmental justice must recognize a special legal and natural relationship of Native Peoples to the U.S. government through treaties, agreements, compacts, and covenants affirming sovereignty and self-determination.

12. Environmental justice affirms the need for urban and rural ecological policies to clean up and rebuild our cities and rural areas in balance with nature, honoring the cultural integrity of all of our communities, and providing fair access for all to the full range of resources.

13. Environmental justice calls for the strict enforcement of principles of informed consent, and a halt to the testing of experimental reproductive and medical procedures and vaccinations on people of color.

14. Environmental justice opposes the destructive operations of multinational corporations.

15. Environmental justice opposes military occupation, repression, and exploitation of lands, peoples, and cultures, and other life forms.

16. Environmental justice calls for the education of present and future generations which emphasizes social and environmental issues, based on our experience and an appreciation of our diverse cultural perspectives.

17. Environmental justice requires that we, as individuals, make personal and consumer choices to consume as little of Mother Earth's resources and to produce as little waste as possible; and make the conscious decision to challenge and re-prioritize our lifestyles to insure the health of the natural world for present and future generations.

Notes

NOTES TO THE INTRODUCTION

1. Or what Brulle (2000) calls "reform environmentalism."

2. Hempel (1999) argues that sustainability advocates come from at least four root approaches: the natural-capital/capital-theory approach of economists; the urban-design approach of land-use planners and architects; the ecosystem-management approach of ecologists and resource managers; and the metropolitan-governance approach of regional policy planners. Brulle (2000) interestingly includes sustainable development under his *conservation* discourse.

3. By "professional" I do not wish to insinuate that smaller, neighborhood groups are unprofessional. My point here is that professional groups are ones with a staff, often with staffers qualified in a cognate discipline.

4. Brulle and Schaefer Caniglia (2000:15) call frame analysis "the study of social movements from a linguistic perspective." They continue that a discursive frame is "the taken-for-granted reality in which a social movement exists. It provides an interpretation of history that defines the origins of the movement, heroic exemplars of the movement, its process of development and future agenda. This narrative provides a group identity for movement organizations and guidance for collective actions" (ibid.). See also Benford and Hunt (1992).

5. I use the word *paradigm* partly in the Kuhnian (1962) sense—as a comprehensive way of seeing the world, as a worldview. However, I also see paradigms as Ritzer (1975:7) does—"a fundamental image of the subject matter" within a discipline. In other words, in this book, paradigms both describe *content* and *worldview.*

6. See http://www.jtalliance.org/docs/aboutjta.html.

7. Carmin and Balser (2002:366) note that "repertoires may consist of *institutional* tactics such as lobbying, litigation, and educational campaigns or *expressive* tactics such as protest, boycotts, and street theatre" (my emphasis).

8. The basic elements of the framework consist of five basic characteristics:

a. Incorporates the principle of the right of all individuals to be protected from environmental degradation

b. Adopts a public health model of prevention (elimination of the threat before harm occurs) as the preferred strategy
c. Shifts the burden of proof to polluters and dischargers who do harm or discriminate or who do not give equal protection to racial and ethnic minorities and other "protected" classes
d. Allows disparate impact and statistical weight, as opposed to "intent" to infer discrimination
e. Redresses disproportionate risk burdens through targeted action and resources. (Bullard 1994:10)

9. Local Agenda 21 was renamed Local Action 21 at the World Summit on Sustainable Development in Johannesburg in September 2002.

10. Local governments are, however, undertaking a wide range of sustainability projects (see ICLEI 2002a). Pressure from NGOs should yield a greater focus on equity and justice.

NOTES TO CHAPTER 1

1. I have called this section "a brief history" because every book with an environmental justice theme seems to start with an expansive treatise on the subject. For fuller histories, see Bullard (1994) and Faber (1998).

2. Cole and Foster (200:19) argue that pinpointing a start date for the movement is impossible because the movement "grew organically out of dozens, even hundreds, of local struggles and events and out of a variety of other social movements."

3. Faber and McCarthy (2003) identify six political movements from which the environmental justice movement has emerged: civil rights, occupational health and safety, indigenous land rights, public health and safety, human rights/solidarity, and social and economic justice.

4. Brulle (2000:213) talks about "the movements for environmental justice," differentiating between the "citizen-worker movement" and the "people of color environmental movement."

5. The Natural Resources Defense Council, Environmental Policy Institute, National Wildlife Federation, Environmental Defense Fund, Izaak Walton League, Sierra Club, National Audubon Society, National Parks and Conservation Association, Wilderness Society, and Friends of the Earth.

6. As Cole and Foster (2001:33) note, however, "because of their backgrounds, these activists often have a distrust for the law and are often experienced in the use of nonlegal strategies, such as protest and other direct action."

7. The Precautionary Principle is hinted at in Bullard's (1994) "environmental justice framework," but it is not fully developed there. The principle was

more fully developed at the Wingspread Conference, which took place January 23–25, 1998, in Racine, Wisconsin.

8. See http://urbanhealthinstitute.jhu.edu/cbpr.html.

9. Much of the work on CBPR is health based, but its usefulness extends beyond health issues.

10. The word "scientist" merely reflects Peggy Shephard's work in environmental health. The word "researcher" could just as easily have been used.

11. This cleavage is not as clear-cut as activist versus academic, since some people are members of both groups. It is, however, a guide based on observations at meetings and conferences.

12. Although the development of suitable claims, together with appropriate framing, is clearly important to the efficacy of social movements (Capek 1993; Benford 1993; Sandweiss 1998; Novotny 2000; and Taylor 2000), it is, according to Taylor (2000), not enough on its own to account for participation. She outlines several other requisites for activism including the development of social and institutional networks (including micro-structural networks); the ability to mobilize resources; the development and exploitation of political opportunities; and the transformation of consciousness and individual behavior implicit in cognitive liberation.

13. Despite its development in 1978, the NEP is still, to my knowledge, the current and most widely quoted sustainability paradigm, despite the term *sustainability* not being in regular usage in 1978. Interestingly, however, Dunlap (2002:336) notes that "because it seeks to emphasize the ecological foundation of human societies we quickly relabeled it 'The New Ecological Paradigm.'" I will use the original meaning of the NEP throughout this book. The Stockholm Environment Institute (2002) has proposed a New Sustainability Paradigm that is still in production.

14. In March 2000, the American Chemistry Council surveyed environmental justice programs in fifty states. At that time, only nine had environment justice policies. By January 2004, according to the American Bar Association (2004:4), "more than 30 states have expressly addressed environmental justice."

15. Text of Executive Order 12898, "Federal Actions to Address Environmental Justice in Minority Populations and Low-Income Populations," Annotated with Proposed Guidance on Terms in the Executive Order, Section 1-1, Implementation.

16. MassGIS EJ Populations maps are available at http://www.state.ma.us/mgis/ej.htm.

17. Faber and Krieg's initial report was embargoed until January 9, 2001. My later citation (2002) is their preferred citation in *Environmental Health Perspectives*.

18. Eady was hired by Robert Durand, then Secretary for Environmental

Affairs, to be Director of Environmental Justice and Brownfields. She was the architect of the Commonwealth's policy.

19. For a detailed look at EJ in the Mystic, see Agyeman and Bryan (2005).

20. Massachusetts Department of Revenue, Municipal Data Bank, http://www.dls.state.ma.us/MDMSTUF/Socioeconomic/Wealth.xls and http://www.dls.state.ma.us/MDMSTUF/Socioeconomic/ComparisonReport.xls.

21. City of Boston, Department of Neighborhood Development, http://www.cityofboston.gov/dnd/PDFs/Profiles/East_Boston_PD_Profile.pdf.

22. The policy represents a balancing act among community desires to go further, business desires to loosen restrictions, and state desires to achieve something workable.

23. On July 23, 2002, Executive Order 438 established the Massachusetts State Sustainability Program. The state's website notes that "this new Program will work to ensure that state government remains in compliance with all environmental laws and regulations, while serving as a model by promoting sustainable practices that reduce the state's environmental impact and save taxpayer dollars" (http://www.state.ma.us/envir/sustainable/default.htm). The predominant orientation of *environmental* sustainability in Massachusetts state policy is notable here.

24. Massachusetts's Sustainable development principles:

1. *Redevelop First.* Support the revitalization of community centers and neighborhoods. Encourage reuse and rehabilitation of existing infrastructure rather than the construction of new infrastructure in undeveloped areas. Give preference to redevelopment of brownfields, preservation and reuse of historic structures and rehabilitation of existing housing and schools.
2. *Concentrate Development.* Support development that is compact, conserves land, integrates uses, and fosters a sense of place. Create walkable districts mixing commercial, civic, cultural, educational and recreational activities with open space and housing for diverse communities.
3. *Be Fair.* Promote equitable sharing of the benefits and burdens of development. Provide technical and strategic support for inclusive community planning to ensure social, economic, and environmental justice. Make regulatory and permitting processes for development clear, transparent, cost-effective, and oriented to encourage smart growth and regional equity.
4. *Restore and Enhance the Environment.* Expand land and water conservation. Protect and restore environmentally sensitive lands, natural resources, wildlife habitats, and cultural and historic landscapes. Increase the quantity, quality and accessibility of open space. Pre-

serve critical habitat and biodiversity. Promote developments that respect and enhance the state's natural resources.

5. *Conserve Natural Resources.* Increase our supply of renewable energy and reduce waste of water, energy and materials. Lead by ex-ample and support conservation strategies, clean power and innovative industries. Construct and promote buildings and infrastructure that use land, energy, water and materials efficiently.
6. *Expand Housing Opportunities.* Support the construction and rehabilitation of housing to meet the needs of people of all abilities, income levels and household types. Coordinate the provision of housing with the location of jobs, transit and services. Foster the development of housing, particularly multifamily, that is compatible with a community's character and vision.
7. *Provide Transportation Choice.* Increase access to transportation options, in all communities, including land- and water-based public transit, bicycling, and walking. Invest strategically in transportation infrastructure to encourage smart growth. Locate new development where a variety of transportation modes can be made available.
8. *Increase Job Opportunities.* Attract businesses with good jobs to locations near housing, infrastructure, water, and transportation options. Expand access to educational and entrepreneurial opportunities. Support the growth of new and existing local businesses.
9. *Foster Sustainable Businesses.* Strengthen sustainable natural resource–based businesses, including agriculture, forestry and fisheries. Strengthen sustainable businesses. Support economic development in industry clusters consistent with regional and local character. Maintain reliable and affordable energy sources and reduce dependence on imported fossil fuels.
10. *Plan Regionally.* Support the development and implementation of local and regional plans that have broad public support and are consistent with these principles. Foster development projects, land and water conservation, transportation and housing that have a regional or multi-community benefit. Consider the long-term costs and benefits to the larger Commonwealth. (http://www.state.ma.us/dhcd/components/housdev/10SDprin.pdf)

NOTES TO CHAPTER 2

1. At the Local Government Summit of the World Summit on Sustainable Development held in Johannesburg in August–September 2002, Local Agenda 21 was renamed Local Action 21. Local Action 21 was launched as a motto for the second decade of Local Agenda 21, a mandate to local authorities world-

wide to move from agenda to action and ensure an accelerated implementation of sustainable development, and to strengthen the LA21 movement of local governments to create sustainable communities and cities while protecting global common goods.

2. The American Planning Association's (APA) "Policy Guide on Planning for Sustainability" (adopted by Chapter Delegate Assembly, April 16, 2000, and ratified by the Board of Directors, April 17, 2000, New York) notes that "planners can therefore play a crucial role in improving the sustainability of communities and the resources that support them." The APA notes that

> planning for sustainability requires a systematic, integrated approach that brings together environmental, economic and social goals and actions directed toward the following four objectives:
>
> 1. Reduce dependence upon fossil fuels, extracted underground metals and minerals.
>
>   Reason: Unchecked, increases of such substances in natural systems will eventually cause concentrations to reach limits—as yet unknown—at which irreversible changes for human health and the environment will occur and life as we know it may not be possible.
> 2. Reduce dependence on chemicals and other manufactured substances that can accumulate in Nature.
>
>   Reason: Same as before.
> 3. Reduce dependence on activities that harm life-sustaining ecosystems.
>
>   Reason: The health and prosperity of humans, communities, and the Earth depend upon the capacity of Nature and its ecosystems to reconcentrate and restructure wastes into new resources.
> 4. Meet the hierarchy of present and future human needs fairly and efficiently.
>
>   Reason: Fair and efficient use of resources in meeting human needs is necessary to achieve social stability and achieve cooperation for achieving the goals of the first three guiding policies.

3. See http://www.sustainablemeasures.com.

4. Examples of indicators in use include racial and ethnic representation in legislature; unemployment rate by ethnicity; occupational distribution of women and minorities; number of homeless people; number and value of business loans in low-income area; percent of jobs that pay a livable wage for a family of two; number or percent of residents receiving welfare assistance.

5. According to McLaren (2003:23), "Chlorinated organic compounds are assessed as an example of the need to eliminate such toxic, persistent and bioaccumulative materials outside of use in closed systems."

6. For a full description of the methodology of environmental space, see McLaren, Bullock, and Yousuf (1998).

7. In 1997, a team of researchers from the United States, Argentina, and the Netherlands, led by Costanza, then of the University of Maryland, put an average price of US$33 trillion a year on "fundamental ecosystem services," which are largely taken for granted because they are free. Because of the nature of the uncertainties, this must be considered a minimum estimate. That is nearly twice the value of the global gross national product (GNP) of US$18 trillion.

8. Hawken et al. (1999:4) describe human capital as "labor and intelligence, culture and organization."

9. Other measures include the Genuine Progress Indicator of John and Clifford Cobb (see chapter 3) and the UN's Human Development Index.

10. The ecological footprint is an accounting tool for ecological resources. Categories of human consumption are translated into areas of productive land required to provide resources and assimilate waste products. The ecological footprint is a measure of how sustainable our lifestyles are. Wackernagel and Rees (1996:15) conclude that "if everybody lived like today's North Americans, it would take at least two additional planet Earths to produce the resources, absorb the wastes, and otherwise maintain life-support."

11. See Warner's (2002) list of cities with sustainability projects, Portney's (2003) list of cities "taking sustainability seriously," Lake's (2000) study, and ICLEI (2002c).

12. There are many guidelines for the development of sustainable communities including the Melbourne Principles for Sustainable Cities, the Hannover Principles, and the Ahwahnee Principles on Smart Growth.

13. In the report, Appendix A, entitled "Definitions and Principle of Sustainable Communities," lists the ISC's Elements of a Sustainable Community and the Ahwahnee Principles on Smart Growth, followed by the Principles of Environmental Justice, on page 47. It is interesting that there seems to have existed among the Task Force the linkage between sustainable communities and environmental justice. This contrasts with Warner's (2002) findings regarding U.S. cities with sustainability initiatives and their non-incorporation of environmental justice.

14. See, for example, http://www.bestpractices.org.

15. Portney's (2003) thirty-four variables used in his Index of Taking Sustainable Cities Seriously:

Sustainable Indicators Project

1. Indicators project active in last five years
2. Indicators progress report in last five years
3. Does indicators project include "action plan" of policies/programs?

"Smart Growth" Activities

4. Eco-industrial park development
5. Cluster or targeted economic development
6. Eco-village project or program
7. Brownfield redevelopment (project or pilot project)

Land Use Planning Programs, Policies, and Zoning

8. Zoning used to delineate environmentally sensitive growth areas
9. Comprehensive land use plan that includes environmental issues
10. Tax incentives for environmentally friendly development

Transportation Planning Programs and Policies

11. Operation of public transit (buses and/or trains)
12. Limits on downtown parking spaces
13. Car pool lanes (diamond lanes)
14. Alternatively fueled city vehicle program
15. Bicycle ridership program

Pollution Prevention and Reduction Efforts

16. Household solid waste recycling
17. Industrial recycling
18. Hazardous waste recycling
19. Air pollution reduction program (i.e., VOC reduction)
20. Recycled product purchasing by city government
21. Superfund site remediation
22. Asbestos abatement program
23. Lead paint abatement program

Energy and Resource Conservation/Efficiency Initiatives

24. Green building program
25. Renewable energy use by city government
26. Energy conservation effort (other than Green building program)
27. Alternative energy offered to consumers (solar, wind, biogas, etc.)
28. Water conservation program

Organization/Administration/Management/Coordination/Governance

29. Single gov/nonprofit agency responsible for implementing sustainability
30. Part of a citywide comprehensive plan
31. Involvement of city/county/metropolitan council
32. Involvement of mayor or chief executive officer
33. Involvement of the business community (e.g., Chamber of Commerce)
34. General public involvement in sustainable cities initiative (public hearings, "visioning" process, neighborhood groups or associations, etc.)

16. Brulle (2000) lists nine discourses of the U.S. environmental movement: manifest destiny, wildlife management, conservation, preservation, reform environmentalism, deep ecology, environmental justice, eco-feminism, and eco-theology. My assertion regarding civic environmentalism focuses on its dominance in policymaking, not in broader environmental thought, which is what Brulle is talking about.

17. Layzer (2002:2) defines civic environmentalism as "using local, collaborative decision-making processes to generate innovative, non-regulatory solutions to a host of environmental problems." This is a narrow-focus approach to civic environmentalism when compared to that of authors such as Shutkin (2000), Roseland (1998), Hempel (1999), and Mazmanian and Kraft (1999).

18. Sabel et al. (1999) could fit in both narrow and broad focus categories. On the one hand, they talk of *backyard environmentalism,* which is a broad conception, but on the other, their focus is still on environmental issues, not on broader economic and social—that is, sustainability—concerns. It is for this reason that I include them under the narrow focus.

19. See http://www.rppi.org; http://www.newenvironmentalism.org.

NOTES TO CHAPTER 3

1. This chapter is based in part on two earlier works: Agyeman, J., Bullard, R., and Evans, B. (2002) "Exploring the Nexus: Bringing Together Sustainability, Environmental Justice and Equity," *Space and Polity,* Vol. 6, No. 1, pp. 77–90; and Evans, T., and Agyeman, J. (2003) "Towards 'Just Sustainability' in Urban Communities: Building Equity Rights with Sustainable Solutions," *Annals of the American Academy of Political and Social Science,* Vol. 590, pp. 35–53.

2. See http://www.sustainable-city.org/Plan/Justice/intro.htm.

3. McLaren (2003:35) argues that "an estimate of $1500 billion is arrived at, based on an increased atmospheric carbon stock of 160 gigatonnes, 80 percent of which is derived from developed countries, which on average over this period account for around one-third of global population, with a value of $20 per tonne of emissions."

4. While the Earth Charter may be a document aimed at national governments, it has applicability at the local level. On July 1, 2000, at Global Cities 21, the ICLEI World Congress for Local Governments, the international membership of ICLEI endorsed the Earth Charter. In addition, for Vermont's Town Meeting 2002, thirty towns had an article on their agendas that read, "Shall the voters of [town] endorse the Earth Charter, and recommend that the Town, the State of Vermont, the United States of America, and the United Nations use the Earth Charter to guide decision-making on issues of local, state, national, and international importance." The twenty-one towns that endorsed the charter are Bethel, Bristol, Bolton, Charlotte, Granby, Hinesburg, Huntington, Isle La

Motte, Lincoln, Marlboro, Marshfield, Middlebury, Monkton, Norwich, Plainfield, Randolph, Ripton, Starksboro, Warren, Weston, and Weybridge.

5. Dobson (1999, 2003) uses the term "environmental sustainability" in all his arguments. He sees sustainability in the environmental sense, rather than my more inclusive sense. This contrasts markedly with Hempel's (1999:43) point, discussed in chapter 2, that "the emerging sustainability ethic may be more interesting for what it implies about politics than for what it promises about ecology."

6. Proposed Principles of Environmental Justice (Boardman et al. 1999):

> 1. Environmental problems are a component of social exclusion and an issue of social justice. Most environmental pollution is unevenly distributed: even in rich countries like Britain, it is normally the poor and disadvantaged who suffer most as a result. Even where the effects are more even, impacts are uneven as the rich can more easily respond.
>
> 2. Communities and individuals should have a right to know and the ability to respond to distributed environmental hazards. This means that government should enthusiastically support and implement the Aarhus Convention, incorporate environmental objectives in area-based regeneration initiatives, work to strengthen participation and fully involve communities in locally based strategies.
>
> 3. A general improvement in the environment will bring disproportionate benefits to the poor and disadvantaged. Other things being equal, this follows from the uneven distribution of impacts. But in practice other factors need to be considered too—because the means as well as the outcomes of improvement will have distributed impacts.
>
> 4. The poor should not be required to pay for clearing up the environmental "mess" caused mainly by the over-consumption of the rich. Moreover, while consumption behavior has to change, the poor—in Britain and elsewhere—should not be denied a decent level of consumption. Quality of life for all is compatible with a convergence of material consumption rates at sustainable levels.
>
> 5. Devising policies to increase quality of life for all demands explicit outcomes and appropriate indicators including an overall measure of quality of life. Within such a policy framework, environmental and equity objectives can be met if both are explicit objectives—accompanied by appropriate measures and indicators.
>
> 6. Government action has a key role in ensuring that the means of environmental improvement are socially just. Existing inequalities, such as income or housing quality, mean that mechanisms designed to change individual or corporate behavior—such as taxes—can be socially regressive if applied in isolation.

7. Policy packages, positively incorporating the goal of socially progressive outcomes, are better at delivering environmental outcomes than isolated policies. Well-designed packages, such as those built around tax reforms and market transformation, have the ability to deliver change more quickly and protect the poor from adverse impacts.

8. Investment is normally preferable to additional income as the most effective way to deal with environmental injustice. Solving environmental problems tends to involve investing capital—for example, in improving housing, or providing public transport—but the poor do not have access to capital. Investment helps the poor to have a better standard of living and quality of life.

9. Environmental modernization through taxes and regulations directed at business and market transformations can be designed to have a largely beneficial impact on poor households. Environmental tax reforms can reduce employment taxes, triggering increased employment, while market transformation policies can include rebates for environmentally efficient goods targeted at the poor.

10. Further research and better monitoring is needed. Improved information on the social distribution of environmental problems is essential for policy-makers, and for the communities seeking to improve their own environment.

7. See McKenzie-Mohr and Smith (1999).

8. Note that the first principle calls for a just sustainability. The idea that just sustainability is a concept that comes from the EJ movement is dealt with in the ACE case study in chapter 5.

NOTES TO CHAPTER 4

1. This chapter is based in part on two earlier works: Agyeman, J., Bullard, R., and Evans, B. (2002) "Exploring the Nexus: Bringing Together Sustainability, Environmental Justice and Equity," *Space and Polity,* Vol. 6, No. 1, pp. 77–90; and Evans, T., and Agyeman, J. (2003) "Towards 'Just Sustainability' in Urban Communities: Building Equity Rights with Sustainable Solutions," *Annals of the American Academy of Political and Social Science,* Vol. 590, pp. 35–53.

2. The choice of land-use planning, solid waste, toxic chemical use, residential energy use, and transportation as sustainability issue categories reflects my need to show, in characteristic areas of concern, how justice and sustainability play out in practice. Other areas could be chosen, but these categories overlap and extend into areas such as housing and employment.

3. Bus Riders Union, http://www.busridersunion.org; New York City Environmental Justice Alliance, http://www.nyceja.org; Alaska Community Action

on Toxics, http://www.akaction.net; Reuse Development Organization, http://www.redo.org; Bethel New Life, http://www.bethelnewlife.org; Green Institute, http://www.greeninstitute.org; Massachusetts Energy Consumers Alliance, http://www.massenergy.com; National Center for Alternative Technology, http://www.ncat.org; Silicon Valley Toxics Coalition, http://www.svtc.org; Toxic Use Reduction Institute, http://www.turi.org; Unity Council, http://www.unitycouncil.org; Urban Ecology, http://www.urbanecology.org; Transportation Alternatives, http://www.transalt.org; Communities for a Better Environment, http://www.cbecal.org.

4. See, for instance, Myron Orfield's *American Metropolitics: The New Suburban Reality* (Washington, DC: Brookings Institution Press, 2002), in which he outlines a comprehensive regional agenda to address urban problems, with solutions for land-use planning from a regional perspective, greater fiscal equity among local governments (with an emphasis on reinvestment in the central cities and older suburbs), and improved governance at the regional level that will help facilitate the development of policies to benefit all types of metropolitan communities.

5. In 1996, the U.S. Environmental Protection Agency's Urban and Economic Development Division launched a series of meetings with hundreds of individuals and organizations in an attempt to build consensus around land-use issues and figure out how to get better information and tools into the hands of local officials, planners, developers, preservationists, environmentalists, and others who were battling sprawl. That year, a broad coalition formally joined hands as the Smart Growth Network (SGN). The SGN came up with ten Smart Growth principles:

1. Mix land uses.
2. Take advantage of compact building design.
3. Create housing opportunities and choices.
4. Create walkable communities.
5. Foster distinctive, attractive communities with a strong sense of place.
6. Preserve open space, farmland, natural beauty, and critical environmental areas.
7. Strengthen and direct development toward existing communities.
8. Provide a variety of transportation choices.
9. Make development decisions predictable, fair, and cost-effective.
10. Encourage community and stakeholder collaboration in development decisions.

NOTES TO CHAPTER 5

1. Interviews—general questions for staff:

    a. What attracted you to working for ACE, and when did you start?
    b. What is your job title, and which program are you working on?
    c. What do you like most, and least about your job?
    d. Looking at ACE's *programs,* from the Asphalt campaign in 1993 to the Safety Net in 2004, do you think they've changed? (*hand out timeline*)
    If so, *how* have they changed?
    Probe: predominantly local to more regional, reactive to proactive, single issue to more systemic, single organization to coalition based
    If so, *why* have they changed?
    Probe: more effective, greater range of skills in coalitions
    e. What tools and techniques do ACE use to increase *community involvement* in its programs?
    Probe: email, word of mouth, or community forums, visioning, etc.
    f. From the Asphalt campaign in 1993 to the Safety Net in 2004, do you think ACE's *community involvement* tools and techniques have changed?
    If so, *how* have they changed?
    Probe: more active techniques
    If so, *why* have they changed?
    Probe: greater level of involvement
    g. List the *campaigning* tools and techniques ACE uses in its programs?
    Probe: Institutional tactics such as lobbying, litigation, educational campaigns or expressive tactics such as protest, boycotts, and street theatre . . . or both, and more?
    h. From the Asphalt campaign in 1993 to the Safety Net in 2004, do you think the *campaigning* tools and techniques have changed?
    If so, *how* have they changed?
    Probe: more attention grabbing
    If so, *why* have they changed?
    Probe: greater clarity of message?
    i. Who are ACE's main coalition partners?
    Probe: How would you characterize these groups: EJ, environmental, sustainability, housing, employment, etc.?
    Are these short-term marriages of convenience or longer-term coalitions?
    j. From the Asphalt campaign in 1993 to the Safety Net in 2004 do you think ACE's coalition partners have changed?

If so, *how* have they changed?
Probe: need complementary skills?
If so, *why* have they changed?
Probe: greater strength of message?

k. You mentioned xxx as a critical moment in ACE's history. Can you tell me why you think it was a critical moment?

Critical Moments List—April 20, 2004

Name	*Suggested by*
Anti-idling March	Julian (as an example)
Klare Allen	Warren
Bill and Charlie Leaving	Khalida
Youth as Organizers (Tata, etc.)	Jodi
2000 Fare Increase Campaign/TRU	Penn
Anti-Idling March/Governor's Meeting	Warren
Asphalt Plant (First Case)	Jodi
Safety Net Campaign	Quita
Dudley Garage/Blair's Lot	Penn
Toxic Tour (New Tool)	Jodi
Penn as Executive Director	Warren
BioLab Campaign	Celina
Membership	Khalida
EJ in the 'Hood & GBEJN	Penn
EJ curriculum	Klare Allen
First Orchard Park Meeting	Penn

2. Interviews—general questions for board members:

a. What attracted you to ACE's board?
b. What is your main program interest at ACE? (*hand out timeline*)
c. Which program do you think has most and least potential at ACE?
d. Looking at ACE's *programs,* from the Asphalt campaign in 1993 to the Safety Net in 2004, do you think they've changed? (*hand out timeline*)
If so, *how* have they changed?
Probe: predominantly local to more regional, reactive to proactive, single issue to more systemic, single organization to coalition based
If so, *why* have they changed?
Probe: more effective, greater range of skills in coalitions
e. what tools and techniques do ACE use to increase *community involvement* in its programs?
Probe: email, word of mouth, or community forums, visioning, etc.

f. From the Asphalt campaign in 1993 to the Safety Net in 2004, do you think ACE's *community involvement* tools and techniques have changed?
If so, *how* have they changed?
Probe: more active techniques
If so, *why* have they changed?
Probe: greater level of involvement

g. List the *campaigning* tools and techniques ACE uses in its programs?
Probe: Institutional tactics such as lobbying, litigation, educational campaigns or expressive tactics such as protest, boycotts, and street theatre . . . or both, and more?

h. From the Asphalt campaign in 1993 to the Safety Net in 2004, do you think the *campaigning* tools and techniques have changed?
If so, *how* have they changed?
Probe: more attention grabbing
If so, *why* have they changed?
Probe: greater clarity of message?

i. Who are ACE's main coalition partners?
Probe: How would you characterize these groups: EJ, environmental, sustainability, housing, employment, etc.?
Are these short-term marriages of convenience or longer-term coalitions?

j. From the Asphalt campaign in 1993 to the Safety Net in 2004 do you think ACE's coalition partners have changed?
If so, *how* have they changed?
Probe: need complementary skills?
If so, *why* have they changed?
Probe: greater strength of message?

3. The word *minority,* while problematic in a number of ways, is the term used in the census.

4. Hoyte stepped down in spring 2005.

5. Jodi Sugerman-Brozan used to work for Save the Harbor/Save the Bay.

NOTES TO CHAPTER 6

1. In spring 2005, Klare Allen left ACE.

References

Acción Ecológica. (1999) Initial Proposal. http://www.deudaecologica.org/c_propuin.html (accessed February 20, 2004).

Ackerman, F., and Mirza, S. (2001) Waste in the Inner City: Asset or Assault? *Local Environment*. Vol. 6, No. 2, pp. 113–120.

Adeola, F. (1994) Environmental Hazards, Health and Racial Inequity in Hazardous Waste Distribution. *Environment and Behavior*. Vol. 26, No. 1, pp. 99–126.

Adeola, F. (2000) Cross National Environmental Injustice and Human Rights Issues: A Review of Evidence from the Developing World. *American Behavioral Scientist*. Vol. 43, No. 4, pp. 686–706.

Adger, N. (2002) Inequality, Environment and Planning. *Environment and Planning A*. Vol. 34, No. 10, pp. 1716–1719.

Agyeman, J. (1990) Black People in a White Landscape: Social and Environmental Justice. *Built Environment*. Vol. 16, No. 3, pp. 232–236.

Agyeman, J. (2000) Environmental Justice: From the Margins to the Mainstream? Report of the Town and Country Planning Association, "Tomorrow" Series, London.

Agyeman, J. (2001) Ethnic Minorities in Britain: Short Change, Systematic Indifference and Sustainable Development. *Journal of Environmental Policy and Planning*. Vol. 3, No. 1, pp. 15–30.

Agyeman, J., and Angus, B. (2003) The Role of Civic Environmentalism in the Pursuit of Sustainable Communities. *Journal of Environmental Planning and Management*. Vol. 46, No. 3, pp. 345–363.

Agyeman, J., and Bryan, D. (2005) Environmental Justice across the Mystic: Bridging Agendas in a Watershed. In Brugge, D., and Hynes, P. (eds.) *Community Research in Environmental Health: Lessons in Science, Advocacy and Ethics*. Aldershot, UK: Ashgate Press.

Agyeman, J., Bullard, R. D., and Evans, B. (2002) Exploring the Nexus: Bringing Together Sustainability, Environmental Justice and Equity. *Space and Polity*, Vol. 6, No. 1, pp. 70–90.

Agyeman, J., Bullard, R. D., and Evans, B. (2003) *Just Sustainabilities: Development in an Unequal World*. London: Earthscan/MIT Press.

Agyeman, J., and Evans, B. (2004) "Just Sustainability": The Emerging Discourse of Environmental Justice in Britain? *Geographical Journal.* Vol. 170, No. 2, pp. 155–164.

Alinsky, S. (1971) *Rules for Radicals: A Pragmatic Primer for Realistic Radicals.* New York: Random House.

Alston, D. (1991) Speech delivered at the First National People of Color Environmental Leadership Summit. Washington, DC, October.

Alternatives for Community and Environment (2002) Five-Year Strategic Plan for 2002–2007.

Alternatives for Community and Environment (2004) Ten-Year Anniversary Video.

Alternatives Press (1996) Community Academy: Reducing Environmental Causes of Asthma. Vol. 3, p. 7.

Alternatives Press (1997 Don't Just Idle! Vol. 4, p. 1.

Alternatives Press (1998) A Tribute to Alternatives for Community and Environment and Its Founders. Vol. 5, p. 7.

Alternatives Press (2000) Riding to Victory: Riders Win Bus Transfers, Weekly Combo Pass. Vol. 7, p. 1.

Alternatives Press (2002) Roxbury Residents Organize a Safety Net. Vol. 9, p. 6.

American Bar Association (2004) Environmental Justice for All: A Fifty State Survey of Legislation, Policies and Initiatives. Chicago: American Bar Association and Hastings College of the Law.

American Chemistry Council (2000) Environmental Justice Programs: 50 State Survey. Washington, DC: American Chemistry Council.

American Planning Association (2000) Policy Guide on Planning for Sustainability. Chicago: American Planning Association.

Anthony, C. (1998) Foreword to Faber, D. (ed.) *The Struggle for Ecological Democracy: Environmental Justice Movements in the United States.* New York: Guilford Press.

Arnstein, S. R. (1969) A Ladder of Citizen Participation. *Journal of the American Planning Association.* Vol. 35, No. 4, pp. 216–224.

Ashman, L., de la Vega, J., Dohan, M., Fisher, A., Hippler, R., and Romain, B. (1993) Seeds of Change: Strategies for Food Security for the Inner City. Unpublished master's client project, Graduate School of Architecture and Planning, University of California, Los Angeles.

Athanasiou, T. (n.d.) Taking Equality Seriously: Environmental Justice as the Future of Environmentalism. http://www.thomhartmann.com/Athanasiou/whole paper.htm (accessed July 12, 2004).

Bachman, W., and Katsev, R. (1984) The Effects of Non-Contingent Free Bus Tickets and Personal Commitment on Urban Bus Ridership. *Transportation Research.* Vol. 16A, No. 2, pp. 103–108.

Barton, H. (ed.) (2000) *Sustainable Communities: The Potential for Eco-Neighborhoods.* London: Earthscan.

Beatley, T., and Manning, K. (1997) *The Ecology of Place*. Washington, DC: Island Press.

Benford, R. (1993) Frame Disputes within the Nuclear Disarmament Movement. *Social Forces*. Vol. 71, No. 3, pp. 677–701.

Benford, R., and Hunt, S. (1992) Dramaturgy and Social Movements: The Social Construction and Communication of Power. *Sociological Inquiry*. Vol. 62, No. 1, pp. 36–55.

Best, J. (1987) Rhetoric in Claims Making. *Social Problems*. Vol. 34, No. 2, pp. 101–121.

Blanding, M. (2004) Fear in the Air. *Boston Magazine*. June, p. 116.

Boardman, B., Bullock, S., and McLaren, D. (1999) *Equity and the Environment: Guidelines for Socially Just Government*. London: Catalyst/Friends of the Earth.

Bossel, H. (1998) *Earth at a Crossroads: Paths to a Sustainable Future*. Cambridge: Cambridge University Press.

Bosselmann, K. (2001) Human Rights and the Environment: Redefining Fundamental Principles? In Gleeson, B., and Low, N. (eds.) *Governing the Environment: Global Problems, Ethics and Democracy*, pp. 118–134. Basingstoke, UK: Palgrave.

Boston Globe (2004) EPA and MBTA Reach $1.3 Million Pollution Settlement. March 10. Available at http://temp.sfgov.org/sfenvironment/aboutus/innovative/pp/sfpp.htm (accessed January 14, 2004).

Boyce, J. K., Klemer, A. R., Templet, P. H., and Willis, C. E. (1999) Power Distribution, the Environment, and Public Health: A State Level Analysis. *Ecological Economics*. Vol. 29, pp. 127–140.

Boyce, J. K., and Pastor, M. (2001) Building Natural Assets. Report of the Political Economy Research Institute/Ford Foundation, University of Massachusetts Amherst.

Brulle, R. (2000) *Agency, Democracy, and Nature: The U.S. Environmental Movement from a Critical Theory Perspective*. Cambridge, MA: MIT Press.

Brulle, R., and Schaefer Caniglia, B. (2000) Money for Nature: A Network Analysis of Foundations and U.S. Environmental Groups. Working Paper and Technical Report Series Number 2000-01, Department of Sociology, University of Notre Dame.

Bryant, B., and Mohai, P. (eds.) (1992) *Race and the Incidence of Environmental Hazards*. Boulder, CO: Westview Press.

Buhrs, T. (2004) Sharing Environmental Space: The Role of Law, Economics and Politics. *Journal of Environmental Planning and Management*. Vol. 47, No. 3, pp. 429–447.

Buitenkamp, M., Venner, H., and Wams, T. (eds.) (1992/1993) *Action Plan Sustainable Netherlands*. Amsterdam: Friends of the Earth Netherlands.

Bullard, R. (1990a) *Dumping in Dixie*. Boulder, CO: Westview Press.

Bullard, R. (1990b) Ecological Inequalities and the New South: Black Communities Under Siege. *Journal of Ethnic Studies*. Vol. 17, No. 4, pp. 101–115.

Bullard, R. (ed.) (1993) *Confronting Environmental Racism: Voices from the Grassroots*. Boston: South End Press.

Bullard, R. (ed.) (1994) *Unequal Protection: Environmental Justice and Communities of Color*. San Francisco: Sierra Club Books.

Bullard, R., and Johnson, G. (eds.) (1997) *Just Transportation*. Washington, DC: Island Press.

Bullard, R., Johnson, G., and Torres, A. (eds.) (2000) *Sprawl City: Race, Politics and Planning in Atlanta*. Washington, DC: Island Press.

Bullard, R., Johnson, G., and Torres, A. (2004) *Highway Robbery: Transportation Racism and New Routes to Equity*. Boston: South End Press.

Burgess, J., Harrison, C., and Filius, P. (1998) Environmental Communication and the Cultural Politics of Environmental Citizenship. *Environment and Planning A*. Vol. 30, pp. 1445–1460.

Campbell, D. (1975) Degrees of Freedom and the Case Study. *Comparative Political Studies*. Vol. 8, pp. 178–185.

Campbell, S. (1996) Green Cities, Growing Cities, Just Cities: Urban Planning and the Contradictions of Sustainable Development. *Journal of the American Planning Association*. Vol. 62, No. 3, pp. 296–312.

Capek, S. (1993) The "Environmental Justice" Frame: A Conceptual Discussion and an Application. *Social Problems*. Vol. 40, No. 1, pp. 5–24.

Carley, M., and Spapens, P. (1997) *Sharing Our World*. London: Earthscan.

Carmin, J., and Balser D. (2002) Selecting Repertoires of Action in Environmental Movement Organizations. *Organization and Environment*. Vol. 15, No. 4, pp. 365–388.

Catton, W., and Dunlap, R. (1978) Environmental Sociology: A New Paradigm. *American Sociologist*. Vol. 13, pp. 41–49.

Christie, I., and Warburton, D. (eds.) (2001) *From Here to Sustainability: Politics in the Real World*. London: Earthscan.

Christoff, P. (1996) Ecological Modernization, Ecological Modernities. *Environmental Politics*. Vol. 5, No. 3, pp. 476–500.

City of San Francisco (2003) SF Precautionary Principle Ordinance. Available at http://temp.sfgov.org/sfenvironment/aboutus/innovative/pp/sfpp.htm (accessed January 14, 2004).

Cobb, C., Halstead, T., and Rowe, J. (1995) If the GDP Is Up, Why Is America Down? *Atlantic Monthly*. Vol. 659, pp. 59–78.

Cole, L., and Foster, S. (2001) *From the Ground Up: Environmental Racism and the Rise of the Environmental Justice Movement*. New York: NYU Press.

Commonwealth of Massachusetts (2002) Environmental Justice Policy. Boston: State House.

Commonwealth of Massachusetts (2004) Massachusetts Climate Protection Plan. Boston: State House, Executive Department.

Congressional Black Caucus Foundation Inc. (2004) African Americans and Climate Change: An Unequal Burden. Washington, DC: CBCF.

Conservation Law Foundation (1998) *City Routes, City Rights: Building Livable Neighborhoods and Environmental Justice by Fixing Transportation.* Boston: Conservation Law Foundation.

Costanza, R. (1997) The Value of the World's Ecosystem Services and Natural Capital. *Nature.* Vol. 387, pp. 253–260.

Costi, A. (1998) Environmental Justice and Sustainable Development in Central and Eastern Europe. *European Environment.* Vol. 8, pp. 107–112.

Cotgrove, S., and Duff, A. (1980) Environmentalism, Middle Class Radicalism and Politics. *Sociology Review.* Vol. 28, pp. 335–351.

Cutter, S. (1995) Race, Class and Environmental Justice. *Progress in Geography.* Vol. 19, No. 1, pp. 111–122.

Daly, H., and Cobb, J. (1989) *For the Common Good: Redirecting the Economy toward Community, the Environment, and a Sustainable Future.* Boston: Beacon Press.

Delgado, G. (1999) The Real Test Is Race. *Colorlines.* Vol. 2, No. 3, available at http://www.arc.org/C_Lines/CLArchive/story2_3_02.html (accessed January 27, 2004).

Department of Neighborhood Development (2002) East Boston Data Profile. Available at http://www.cityofboston.gov/dnd/PDFs/Profiles/East_Boston_PD_Profile.pdf (accessed February 4, 2004).

Department of Transport, Environment and the Regions (DETR) (1998) Sustainable Local Communities for the 21st Century. London: DETR.

Devall, W. (1970) Conservation: An Upper Middle Class Social Movement: A Replication. *Journal of Leisure Research.* Vol. 2, No. 2, pp. 123–126.

Dobson, A. (1999) *Justice and the Environment: Conceptions of Environmental Sustainability and Dimensions of Social Justice.* Oxford: Oxford University Press.

Dobson, A. (2003) Social Justice and Environmental Sustainability: Ne'er the Twain Shall Meet? In Agyeman, J., Bullard, R. D., Evans, B. (eds.) *Just Sustainabilities: Development in an Unequal World.* London. Earthscan/MIT Press.

Dockery, D. W., Arden Pope III, C., Xu, X., Spengler, J. D., Ware, J. H., Fay, M. E., Ferris Jr., B. G., and Speizer, F. E. (1993) An Association between Air Pollution and Mortality in Six U.S. Cities. *New England Journal of Medicine.* Vol. 329, No. 24, pp. 1753–1759.

Dryzek, J. (1990) Green Reason: Communicative Ethics and the Biosphere. *Environmental Ethics.* Vol. 12, pp. 195–210.

Duany, A., Plater-Zyberk, E., and Speck, J. (2000) *Suburban Nation: The Rise of Sprawl and Decline of the American Dream.* New York: North Point Press.

Dunion, K., and Scandrett, E. (2003) The Campaign for Environmental Justice in Scotland. In Agyeman, J., Bullard, R. D., and Evans, B. (eds.) *Just Sustainabilities: Development in an Unequal World.* London: Earthscan/MIT Press.

Dunlap, R. (2002) Paradigms, Theories and Environmental Sociology. In Dunlap, R., Buttel, F., Dickens, P., and Gijswijt, A. (eds.) *Sociological Theory and the Environment. Classical Foundations, Contemporary Insights.* Lanham, MD: Rowman and Littlefield.

Dunlap, R., and Catton, W. (1979) Environmental Sociology. *Annual Review of Sociology.* Vol. 5, pp. 243–273.

Eady, V. (2003) Environmental Justice in State Policy Decisions. In Agyeman, J., Bullard, R. D., and Evans, B. (eds.) *Just Sustainabilities: Development in an Unequal World.* London. Earthscan/MIT Press.

Earth Council (2000) The Earth Charter. Costa Rica: Earth Council.

Economic and Social Research Council (2001) Environmental Justice: Rights and Means to a Healthy Environment. London: ESRC.

Edwards, M. (1992) Sustainability and People of Color. *EPA Journal.* Vol. 18, No. 4, pp. 50–52.

Entman, R. (1993) Framing: Toward Clarification of a Fractured Paradigm. *Journal of Communication.* Vol. 43, No. 4, pp. 51–58.

Environmental Careers Organization (1992) Beyond the Green: Redefining and Diversifying the Environmental Movement. Boston: ECO.

Environmental Justice Resource Center (1997) Healthy and Sustainable Communities: Building Model Partnerships for the 21st Century. Atlanta: EJRC.

Environmental Law Institute (1999) Sustainability in Practice. Washington, DC: ELI.

Environmental Protection Agency (1996) Environmental Justice, Urban Revitalization and Brownfields: The Search for Signs of Hope: A Report on the Public Dialogues on Urban Revitalization and Brownfields: Envisioning Healthy and Sustainable Communities. Washington, DC: EPA.

Environmental Protection Agency (1997) Community-Based Environmental Protection: A Resource Book for Protecting Ecosystems and Communities. Washington, DC: EPA, 230-B-96.

Environmental Protection Agency (2004) You Hold the Key—What You Should Know about Bus Idling. Educational Pamphlet.

Enzensberger, H. (1979) A Critique of Political Ecology. In Cockburn, A., and Ridgeway, J. (eds.) *Political Ecology.* New York: New York Times Books.

Epstein, B. (1997) The Environmental Justice/Toxics Movement: Politics of Race and Gender. *Capitalism, Nature and Socialism.* Vol. 8, No. 3, pp. 63–87.

Evans, S., and Boyte, H. (1986) *Free Spaces: The Source of Democratic Change in America.* New York: Harper and Row.

Faber, D. (ed.) (1998) *The Struggle for Ecological Democracy: Environmental Justice Movements in the United States.* New York: Guilford Press.

Faber D., and Krieg E. (2002) Unequal Exposure to Ecological Hazards: Environmental Injustices in the Commonwealth of Massachusetts. *Environmental Health Perspectives Supplements.* Vol. 110, No. 2, pp. 277–288.

Faber, D., and McCarthy, D. (2003) Neo-Liberalism, Globalization and the Struggle for Ecological Democracy. In Agyeman, J., Bullard, R. D., and Evans, B. (eds.) *Just Sustainabilities: Development in an Unequal World.* London: Earthscan/MIT Press.

Fischer, F. (2002) Not in My Backyard: Risk Assessment and the Politics of Cultural Rationality. Chapter in *Citizens, Experts, and the Environment.* Durham, NC: Duke University Press.

Foreman, C. (1998). *The Promise and Perils of Environmental Justice.* Washington, DC: Brookings Institution Press.

Forkenbrock, D. J., and Schweitzer, L. A. (1999) Environmental Justice in Transportation Planning. *Journal of the American Planning Association.* Vol. 65, No. 1, pp. 96–111.

Foster, S. (2002) Environmental Justice in an Era of Devolved Collaboration. *Harvard Environmental Law Review.* Vol. 26, pp. 459–497.

Foster, S. (2003) From Harlem to Havana: Sustainable Urban Development. *Tulane Environmental Law Journal.* Vol. 16, pp. 783–808.

Freeman, C., Littlewood, S., and Whitney, D. (1996) Local Government and Emerging Models of Participation in the Local Agenda 21 Process. *Journal of Environmental Planning and Management.* Vol. 39, No. 1, pp. 65–78.

Freire, P. (1993) *Pedagogy of the Oppressed.* New York: Continuum Books.

Friedland, L., and Sirianni, C. (1995). Civic Environmentalism. Civic Practices Network, available at http://www.cpn.org/imagemaps/topicon.map?333,242 (accessed July 20, 2001).

Friends of the Earth England, Wales, and Northern Ireland (n.d.) Tomorrow's World. Available at http://www.foe.co.uk/campaigns/sustainable_development/publications/tworld/summary.html (accessed January 30, 2004).

Friends of the Earth Scotland (1999) Press Release: FoE Issue Challenge to Scottish Parliament. January 23, 1999. Edinburgh: Friends of the Earth Scotland.

Friends of the Earth Scotland (2000) The Campaign for Environmental Justice. Edinburgh: Friends of the Earth Scotland.

Gaines, R., and Micklewright, S. (1988) Unfair Brother. *Urban Wildlife.* Vol. 1, No. 3, pp. 37–38.

Gallopin, G., Hammond, A., Raskin, P., and Swart, R. (1997) Branch Points: Global Scenarios and Human Choice. PoleStar Series Report No. 7. Stockholm: Stockholm Environment Institute.

Gedicks, A. (1998) Racism and Resource Colonization. In Faber, D. (ed.) *The*

Struggle for Ecological Democracy: Environmental Justice Movements in the United States. New York: Guilford Press.

Geiser, K. (2001) *Materials Matter: Toward a Sustainable Materials Policy.* Cambridge, MA: MIT Press.

Geiser, K., and Waneck, G. (1994) PCBs and Warren County. In Bullard, R. (ed.) *Unequal Protection: Environmental Justice and Communities of Color.* San Francisco: Sierra Club Books.

General Accounting Office (1983) Siting of Hazardous Waste Landfills and Their Correlation with Racial and Economic Status of Surrounding Communities. Washington, DC: GPO.

Gershon, D., and Gilman, R. (1990) *Household Ecoteam Workbook.* Woodstock, NY: Global Action Plan.

Glasmeier, A., and Farrigan, T. (2003) Poverty, Sustainability, and the Culture of Despair: Can Sustainable Development Strategies Support Poverty Alleviation in America's Most Environmentally Challenged Communities? *Annals of the American Academy of Political and Social Sciences.* Vol. 590, pp. 131–149.

Goldman, B. (1993) Not Just Prosperity: Achieving Sustainability with Environmental Justice. Report. Washington, DC: National Wildlife Federation.

Goldman, B. (1996) The Future of Environmental Justice. *Antipode.* Vol. 28, No. 3, pp. 122–141.

Goldman, B. (2000) An Environmental Justice Paradigm for Risk Assessment. *Human and Ecological Risk Assessment.* Vol. 6, No. 6, pp. 541–548.

Gottlieb, R. (1993) *Forcing the Spring: The Transformation of the Environmental Movement.* Washington, DC: Island Press.

Gottlieb, R., and Fisher, A. (1996). First Feed the Face: Environmental Justice and Community Food Security. *Antipode.* Vol. 28, No. 2, pp. 193–203.

Gould, K., Lewis, T., and Roberts, T. (2004) Blue-Green Coalitions: Constraints and Possibilities in the Post 9-11 Political Environment. *Journal of World-Systems Research.* Vol. 10, No. 1, Winter, pp. 90–116.

Gouldson, A., and Murphy, J. (1997) Ecological Modernization and the Restructuring of Industrial Economies. In Jacobs, M. (ed.) *Greening the Millennium?* Oxford: Blackwell.

Gross, E. (1997) Sustainable Communities: Working across Disciplines. In Environmental Justice Resource Center, Healthy and Sustainable Communities: Building Model Partnerships for the 21st Century. Atlanta: EJRC.

Guha, R., and Martinez-Alier, J. (1997) *Varieties of Environmentalism: Essays North and South.* London: Earthscan.

Hajer, M. (1995) *The Politics of Environmental Discourse: Ecological Modernization and the Policy Process.* Oxford: Oxford University Press.

Hannigan, J. (1995) *Environmental Sociology.* London: Routledge.

Harner, J., Warner, K., Pierce, J., and Huber, T. (2002) Urban Environmental Justice Indices. *Professional Geographer.* Vol. 54, No. 3, pp. 318–331.

Harris, J., Wise, T., Gallagher, K., and Goodwin, N. (2001) *A Survey of Sustainable Development: Social and Economic Dimensions.* Washington, DC: Island Press.

Harvey, D. (1996) *Justice, Nature and the Geography of Difference.* Oxford: Blackwell.

Haughton, G. (1999) Environmental Justice and the Sustainable City. In Satterthwaite, D. (ed.) *Sustainable Cities.* London: Earthscan.

Hawken, P. (1993) *The Ecology of Commerce: A Declaration of Sustainability.* New York: Harper Business.

Hawken, P., Lovins, A., and Lovins, H. (1999) N*atural Capitalism.* Boston: Back Bay Books.

Heidelberg City Development Plan (1999). Available at http://www.heidelberg.de/stadtentwicklung/step2010/english/stepka9e.htm#Ziel02 (accessed July 7, 2004).

Heiman, M. (1996) Race, Waste, and Class: New Perspectives on Environmental Justice. *Antipode.* Vol. 28, No. 2, pp. 111–121.

Heinrich Boll Foundation (2002a) Sustainability and Justice: A Political North-South Dialogue. Berlin: Heinrich Boll Foundation.

Heinrich Boll Foundation (2002b) The Jo'burg Memo: Fairness in a Fragile World. Berlin: Heinrich Boll Foundation.

Hempel, L. C. (1999) Conceptual and Analytical Challenges in Building Sustainable Communities. In Mazmanian, D. A., and Kraft, M. E. (eds.) *Towards Sustainable Communities: Transition and Transformations in Environmental Policy,* pp. 43–74. Cambridge, MA: MIT Press.

Holmes, T., and Scoones, I. (2000) Participatory Environmental Policy Processes: Experiences from North and South. Working Paper. Brighton: Institute of Development Studies, University of Sussex.

Hynes, H. P. (2003) The Chelsea River: Democratizing Access to Nature in a World of Cities. In Boyce, J. K., and Shelley, B. G. (eds.) *Natural Assets: Democratizing Environmental Ownership.* Washington, DC: Island Press.

International Climate Justice Network (2002) Press Release. Johannesburg, August 29.

International Council for Local Environmental Initiatives (2002a) Second Local Agenda 21 Survey. New York: United Nations Commission on Sustainable Development.

International Council for Local Environmental Initiatives (2002b) CCP Five-Milestone Framework. Available at http://www.iclei.org/ccp/five_milestones.htm.

International Council for Local Environmental Initiatives (2002c) Communities

21: A U.S. Local Agenda 21 Initiative. Available at http://www.iclei.org/us/c21_handbook.pdf.

International Union for the Conservation of Nature (1991) *Caring for the Earth.* London: Earthscan.

Jacobs, M. (1991) *The Green Economy: Environment, Sustainable Development and the Politics of the Future.* London: Pluto Press.

Jacobs, M. (1999) Sustainable Development: A Contested Concept. In Dobson, A. (ed.) *Fairness and Futurity: Essays on Environmental Sustainability and Social Justice.* Oxford: Oxford University Press.

John, D. (1994) *Civic Environmentalism.* Washington, DC: Congressional Quarterly Press.

John, D., and Mlay, M. (1999) Community-Based Environmental Protection: Encouraging Civic Environmentalism. In Sexton, K., et al. (eds.) *Better Environmental Decisions: Strategies for Governments, Businesses and Communities.* Washington, DC: Island Press.

Johns Hopkins Urban Health Institute (2003) What Is Community-Based Participatory Research? Available at http://urbanhealthinstitute.jhu.edu/cbpr.html (accessed April 6, 2003).

Kelly, D., and Becker, B. (2000) *Community Planning: An Introduction to the Comprehensive Plan.* Washington, DC: Island Press.

Kirshen, P., Durant, J., and Perez, G. (2000) Water Resource Management in the Mystic River II: University and Community Collaboration through Service Learning and Active Citizenship. Watershed Management 2000, American Society of Civil Engineers, June 21–23.

Kline, E. (1995) Sustainable Community Indicators. Report. Medford, MA: Consortium for Regional Sustainability, Tufts University.

Kollmuss, A., and Agyeman, J. (2002) Mind the Gap: Why Do People Act Environmentally and What Are the Barriers to Pro-environmental Behavior? *Environmental Education Research.* Vol. 8, pp. 239–260.

Kothari, S., and Parajuli, P. (1993) No Nature without Social Justice: A Plea for Ecological and Cultural Pluralism in India. In Sachs, W. (ed.) *Global Ecology: A New Arena of Political Conflict.* London: Zed Books.

Krishnan, R., Harris, J., and Goodwin, N. (eds.) (1995) *A Survey of Ecological Economics.* Washington, DC: Island Press.

Kuhn, T. (1962) *The Structure of Scientific Revolutions.* Chicago: University of Chicago Press.

Lafferty, W. (ed.) (2001) *Sustainable Communities in Europe.* London: Earthscan.

Lake, R. (2000) Contradictions at the Local State: Local Implementation of the U.S. Sustainability Agenda in the USA. In Low, N., and Gleeson, B. (eds.) *Consuming Cities: The Urban Environment in the Global Economy after the Rio Declaration.* London: Routledge.

Landy, M. K., Susman, M. M., and Knopman, D. S. (1999) Civic Environmentalism in Action: A Field Guide to Regional and Local Initiatives. Report. Washington, DC: Progressive Policy Institute, Center for Innovation and the Environment.

Lavelle, M., and Coyle M. (eds.) (1992) Unequal Protection: The Racial Divide in Environmental Law (Special Supplement). *National Law Journal.* Vol. 15, pp. 52–54.

Layzer, J. (2002) Science, Citizen Involvement, and Collaborative Environmental Policymaking. An unpublished research proposal for Tufts University.

Levy, J., Houseman, E. A., Spengler, J., Loh, P., and Ryan, L. (2001) Fine Particulate Matter and Polycyclic Aromatic Hydrocarbon Concentration Patterns in Roxbury, Massachusetts: A Community-Based GIS Analysis. *Environmental Health Perspectives.* Vol. 109, No. 4, pp. 341–347.

Lichterman, P. (1995) Piecing Together Multicultural Community: Cultural Differences in Community Building among Grass-Roots Environmentalists. *Social Problems.* Vol. 42, No. 4, pp. 513–534.

Longo, P. (1998) Environmental Injustices and Traditional Environmental Organizations. In Camacho, D. E. (ed.) *Environmental Injustices, Political Struggles: Race, Class and the Environment.* Durham, NC: Duke University Press.

Low, N., and Gleeson, B. (1998) *Justice, Society and Nature: An Exploration of Political Ecology.* London: Routledge.

Lowe, P., and Goyder, J. (1983) *Environmental Groups in Politics.* London: Allen and Unwin.

Marcell, K., Agyeman, J., and Rappaport, A. (2004) Cooling the Campus: A Pilot Study Using Social Marketing Methods to Reduce Electricity Use at Tufts University. *International Journal of Sustainability in Higher Education.* Vol. 5, No. 2, pp. 169–189.

Marin, J., and Terrell, B. (2003) Light Rail Makes More Sense for Washington St. Corridor. *Boston Herald.* February 16.

Marsh, D. (2003) Promise and Challenge: Achieving Regional Equity in Boston. Oakland, CA: PolicyLink.

Massachusetts Bay Transportation Authority (2003) MBTA Fare Policy Commitments. MBTA Board of Directors Resolution, December 12.

Massachusetts Senate Committee on Post Audit and Oversight (2002) Attacking Asthma: Combating an Epidemic among Our Children. Senate No. 2505, December. Available at http://www.mass.gov/legis/senate/asthma.htm (accessed June 12, 2004).

Massachusetts State Government (n.d.) State Sustainability. Available at http://www.state.ma.us/envir/sustainable/default.htm (accessed February 11, 2004).

Massey, D., and Denton, N. (eds.) (1993) *American Apartheid and the Making of the Underclass.* Cambridge, MA: Harvard University Press.

Mazmanian, D. A., and Kraft, M. E. (eds.) (1999) *Towards Sustainable Com-*

munities: Transition and Transformations in Environmental Policy. Cambridge, MA: MIT Press.

McAdam, D., McCarthy, J., and Zald, M. (1996) Introduction to McAdam, D., McCarthy, J., and Zald, M. (eds.) *Comparative Perspectives on Social Movements: Political Opportunities, Mobilizing Structures, and Cultural Framings.* Cambridge: Cambridge University Press.

McDonough, W. (1992) The Hannover Principles. Prepared for EXPO 2002, Hannover, Germany.

McGranahan, G., and Satterthwaite, D. (2000) Environmental Health or Ecological Sustainability? Reconciling the Brown and Green Agendas in Urban Development. In Pugh, C. (ed.) *Sustainable Cities in Developing Countries.* London: Earthscan.

McKenzie-Mohr, D., and Smith, W. (1999) *Fostering Sustainable Behavior: An Introduction to Community-Based Social Marketing.* Gabriola Island, BC: New Society Publishers.

McLaren, D. (2003) Environmental Space, Equity and the Ecological Debt. In Agyeman, J., Bullard, R. D., and Evans, B. (eds.) *Just Sustainabilities: Development in an Unequal World.* London: Earthscan/MIT Press.

McLaren, D., Bullock, S., and Yousuf, N. (1998) *Tomorrow's World: Britain's Share in a Sustainable Future.* London: Earthscan.

McNaghten, P., and Urry, J. (1998) *Contested Natures.* London: Sage.

Medoff, P., and Sklar, H. (1994) *Streets of Hope: The Fall and Rise of an Urban Neighborhood.* Boston: South End Press.

Middleton, N., and O'Keefe, P. (2001) *Redefining Sustainable Development.* London: Pluto Press.

Milbrath, L. (1984) *Environmentalists: Vanguards for a New Society.* Albany: SUNY Press.

Milbrath, L. (1989) *Envisioning a Sustainable Society: Learning Our Way Out.* Albany: SUNY Press.

Mohai, P. (2003) Dispelling Old Myths: African American Concern for the Environment. *Environment.* Vol. 45, No. 5, pp. 10–26.

Mohai, P., and Bryant, B. (1992) Environmental Injustice: Weighing Race and Class as Factors in the Distribution of Environmental Hazards. *University of Colorado Law Review.* No. 63, pp. 921–932.

Mohai, P., and Bryant, B. (1998) Is There a "Race" Effect on Concern for Environmental Quality? *Public Opinion Quarterly.* Vol. 62, No. 4, pp. 475–505.

Morello-Frosch, R. (1997) Environmental Justice and California's "Riskscape": The Distribution of Air Toxics and Associated Cancer and Non-Cancer Risks among Diverse Communities. Unpublished dissertation, Department of Health Sciences, University of California, Berkeley.

Mouffe, C. (1993) *The Return of the Political.* London: Verso.

National Academy of Public Administration (2001a) Environmental Justice in

EPA Permitting: Reducing Pollution in High-Risk Communities Is Integral to the Agency's Mission. Washington, DC: NAPA.

National Academy of Public Administration (2001b) Models for Change: Efforts by Four States to Address Environmental Justice. Washington, DC: NAPA.

National Academy of Public Administration (2003) Addressing Community Concerns: How Environmental Justice Relates to Land Use Planning and Zoning. Washington, DC: NAPA.

National Academy of Sciences (1999) *Our Common Journey: A Transition toward Sustainability.* Washington, DC: National Academy Press.

Newman, P., and Kenworthy, J. (1999) *Sustainability and Cities: Overcoming Automobile Dependence.* Washington, DC: Island Press.

New York Times (2004) The Real Cost of Gas. May 28, p. A9.

Novotny, P. (2000) *Where We Live, Work and Play.* Westport, CT: Praeger.

Office for Commonwealth Development (2004) OCD website, http://www.mass.gov/ocd (accessed August 7, 2004).

Office of the Inspector General (2004) EPA Needs to Consistently Implement the Intent of the Executive Order on Environmental Justice. Washington, DC: U.S. Environmental Protection Agency, March 1.

Orfield, M. (1997) *Metropolitics: A Regional Agenda for Community and Stability.* Washington, DC: Brookings Institution Press.

O'Riordan, T. (1970) *Environmentalism.* London: Pion.

Perez, G., Durant, J., and Senn, D. (2002) Don't Flush When It Rains? *Cambridge Chronicle.* April 30.

Perfecto, I. (1995) Sustainable Agriculture Embedded in a Global Sustainable Future: Agriculture in the United States and Cuba. In Bryant, B. (ed.) *Environmental Justice: Issues, Policies and Solutions.* Washington, DC: Island Press.

Pitcoff, W. (1999) Sustaining Community Power: Gregg Watson, Executive Director of DSNI, on Sustainable Development and Community Revitalization. *Shelterforce: The Journal of Affordable Housing and Community Building.* Vol. 103, pp. 20–23.

Plough, A., and Krimsky, S. (1987) The Emergence of Risk Communication Studies: Social and Political Context. *Science, Technology and Human Values.* Vol. 12, No. 3–4, pp. 4–10.

Polese, M., and Stren, R. (2000) The New Sociocultural Dynamics of Cities. In Polese, M., and Stren, R. (eds.) *The Social Sustainability of Cities: Diversity and the Management of Change.* Toronto: University of Toronto Press.

Popatchuk, W. (2002) Moving from Collaborative Processes to Collaborative Communities. Draft report prepared for the Association of Collegiate Schools of Planning Conference, November.

Portney, K. (2003) *Taking Sustainable Cities Seriously: Economic Development, the Environment, and Quality of Life in American Cities.* Cambridge, MA: MIT Press.

President's Council on Sustainable Development (1996) Sustainable America: A New Consensus. Washington, DC: GPO.

President's Council on Sustainable Development (1997) Sustainable Communities Task Force Report. Washington, DC: GPO.

Prugh, T., Costanza, R., and Daly, H. (2000) *The Local Politics of Global Sustainability.* Washington, DC: Island Press.

Pugh, C. (ed.) (2000) *Sustainable Cities in Developing Countries.* London: Earthscan.

Pulido, L. (1994) People of Color, Identity Politics, and the Environmental Justice Movement. Unpublished manuscript, Geography Department, University of Southern California.

Pulido, L. (1996a) *Environmentalism and Economic Justice: Two Chicano Struggles in the Southwest.* Tucson: University of Arizona Press.

Pulido, L. (1996b) A Critical Review of the Methodology of Environmental Racism Research. *Antipode.* Vol. 28, No. 2, pp. 142–159.

Rabe, B. (1999) Sustainability in a Regional Context: The Case of the Great Lakes Basin. In Mazmanian, D. A., and Kraft, M. E. (eds.) *Towards Sustainable Communities: Transition and Transformations in Environmental Policy.* Cambridge, MA: MIT Press.

Rabin, J. (1999a) Judge Hears Debate on Bus Crowds. *Los Angeles Times.* July 20, p. B1.

Rabin, J. (1999b) MTA Board Agrees to Buy 297 Buses Ordered by Court. *Los Angeles Times.* September 30, Metro sec., p. B5.

Rachel's Environment and Health News (1998) Clean Production Part 1. May 13, No. 650.

Rachel's Environment and Health News (1999) The Precautionary Principle. February 19, No. 586.

Raffensperger, C., and Tickner, J. (eds.) (1999) *Protecting Public Health and the Environment: Implementing the Precautionary Principle.* Washington, DC: Island Press.

Raskin, P., Gallopin, G., Gutman, P., Hammond, A., and Swart, R. (1998) Bending the Curve: Toward Global Sustainability. PoleStar Series Report No. 8. Boston: Stockholm Environment Institute.

Redefining Progress (2002) The Johannesburg Summit 2002: A Call for Action. Available at http://www.rprogress.org/newopinion/letters/020222_bush call.html (accessed January 14, 2004).

Redefining Progress (2004) Redefining Progress E-Newsletter. Vol. 2, No. 2, available at http://redefiningprogress.org/newsletter/02-04newsletter/index.html.

Rees, W. E. (1995) Achieving Sustainability: Reform or Transformation? *Journal of Planning Literature.* Vol. 9, No. 4, pp. 343–361.

Renn, O., Webler, T., and Wiedermann, P. (eds.) (1995) *Fairness and Competence in Citizen Participation.* Dordrecht: Kluwer.

Ritzer, G. (1975) *Sociology: A Multiple Paradigm Science.* Boston: Allyn and Bacon.

Roberts, P. (2003) Sustainable Development and Social Justice: Spatial Priorities and Mechanisms for Delivery. *Sociological Inquiry.* Vol. 73, No. 2, pp. 228–244.

Robra, M. (2002) Justice—the Heart of Sustainability: "Talking Points" on the World Summit on Sustainable Development: An Introduction. *Ecumenical Review.* Vol. 54, No. 3, pp. 271–278.

Roseland, M. (ed.) (1997) *Eco-City Dimensions.* Gabriola Island, BC: New Society Publishers.

Roseland, M. (1998) *Toward Sustainable Communities: Resources for Citizens and Their Governments.* Gabriola Island, BC: New Society Publishers.

Ruhl, J. B. (1999) The Co-Evolution of Sustainable Development and Environmental Justice: Cooperation, Then Competition, Then Conflict. *Duke Environmental Law and Policy Forum.* Vol. 9, No. 2, pp. 161–185.

Sabel, C., Fung, A., and Karkkainen, B. (1999). Beyond Backyard Environmentalism: How Communities Are Quietly Refashioning Environmental Regulation. *Boston Review.* Vol. 1, No. 12, available at http://www.bostonreview.mit.edu/BR24.5/sabel.html.

Sachs, A. (1995) Eco-Justice: Linking Human Rights and the Environment. Worldwatch Paper 127.

Sandweiss, S. (1998) The Social Construction of Environmental Justice. In Camacho, D. E. (ed.) *Environmental Injustices, Political Struggles: Race, Class and the Environment.* Durham, NC: Duke University Press.

Satterthwaite, D. (1999) *The Earthscan Reader in Sustainable Cities.* London: Earthscan.

Schlosberg, D. (1999) *Environmental Justice and the New Pluralism: The Challenge of Difference for Environmentalism.* Oxford: Oxford University Press.

Schnaiberg, A., Weinberg, A., and Pellow, D. (2001) Markets and Politics in Urban Recycling: A Tale of Two Cities. Working Paper WP-01-03. Chicago: Northwestern University.

Seabrook, J. (1994) Consumerism and Happiness. *Ethical Consumer.* Vol. 27, January, pp. 12–23.

Selman, P., and Parker, J. (1997) Citizenship, Civicness and Social Capital in Local Agenda 21. *Local Environment.* Vol. 2, No. 2, pp. 171–184.

Shabecoff, P. (1990) Environmental Groups Told They Are Racists in Hiring. *New York Times.* February 1, p. A1.

Shellenberger, M., and Nordhaus, T. (2004) *The Death of Environmentalism: Global Warming Politics in a Post Environmental World.* San Francisco: Breakthrough Institute.

Shepard, P., Northridge, M., Prakas, S., and Stover, G. (2002) Advancing Environmental Justice through Community Based Participatory Research. *En-*

vironmental Health Perspectives Supplement. Vol. 110, No. 2, pp. 139–140.

Shutkin, W. A. (2000). *The Land That Could Be: Environmentalism and Democracy in the Twenty-First Century.* Cambridge, MA: MIT Press.

Smith, D. (2003) *Deliberative Democracy and the Environment.* London: Routledge.

Snow, D., and Benford, R. (1992) Master Frames and Cycles of Protest. In Morris, A., and Mueller, C. (eds.) *Frontiers in Social Movement Theory.* New Haven, CT: Yale University Press.

Southey, S. (2001) Accelerating Sustainability: From Agenda to Action. *Local Environment.* Vol. 6, No. 4, pp. 483–489.

Spangenberg, J. (1995) *Towards a Sustainable Europe.* Brussels: Friends of the Earth Europe.

Spector, M., and Kitsuse, J. (1973) Social Problems: A Reformulation. *Social Problems.* Vol. 20, pp. 145–159.

Stockholm Environment Institute (2002) Great Transition: The Promise and Lure of the Times Ahead. Boston: Stockholm Environment Institute.

Stern, P., and Fineberg, H. (eds.) (1996) *Understanding Risk: Informing Decisions in a Democratic Society.* Washington, DC: National Academy Press.

Study Circles Resource Center (2004). What Is a Study Center? http://www.studycircles.org/pages/what.html (accessed July 19, 2004).

Stutz, J., and Mintzer E. (2003) Affluence and Well-Being, Draft 10. Boston: Tellus Institute.

Summit II Executive Committee (2002) End Environmental Racism Now! Second National People of Color Environmental Leadership Summit.

Sustainable Minnesota (n.d.) Ecological Tax Reform Resources. Available at http://www.me3.org/projects/greentax (accessed February 10, 2004).

Taylor, B. (1992) *Our Limits Transgressed: Environmental Political Thought in America.* Kansas City: University Press of Kansas.

Taylor, D. (1992) Can the Environmental Movement Attract and Maintain the Support of Minorities? In Bryant, B., and Mohai, P. (eds.) *Race and the Incidence of Environmental Hazards: A Time for Discourse.* Boulder, CO: Westview Press.

Taylor, D. (1999) Mobilizing for Environmental Justice in Communities of Color: An Emerging Profile of People of Color Environmental Groups. In Aley, J., Burch, W., Canover, B., and Field, D. (eds.) *Ecosystem Management: Adaptive Strategies for Natural Resources Organizations in the 21st Century.* Philadelphia: Taylor and Francis.

Taylor, D. (2000) The Rise of the Environmental Justice Paradigm. *American Behavioral Scientist.* Vol. 43, No. 4, pp. 508–580.

Timmons Roberts, J., and Toffolon-Weiss, M. (2001) *Chronicles from the Environmental Justice Frontline.* Cambridge: Cambridge University Press.

Torras, M., and Boyce, J. K. (1998) Income, Inequality and Pollution: A Reassessment of the Environmental Kuznets Curve. *Ecological Economics*. Vol. 25, pp. 147–160.

United Nations Environment Program (2002) The Melbourne Principles for Sustainable Cities. Integrative Management Series No 1. Osaka: United Nations Environment Program.

United States Commission on Civil Rights (2003) Not in My Backyard: Executive Order 12,898 and Title VI as Tools for Achieving Environmental Justice. Washington, DC: USCCR.

Urban Habitat Program (1995) Sustainability and Justice: A Message to the President's Council on Sustainable Development. San Francisco: Urban Habitat Program.

U.S. Agency for International Development (2002) Working for a Sustainable World: U.S. Government Initiatives to Promote Sustainable Development. Washington, DC: USAID.

U.S. Census (2000) Roxbury Data Profile. Available at http://www.cityofboston.gov/DND/PDFs/Profiles/Roxbury_PD_Profile.pdf (accessed June 12, 2004).

Veenhoven, R. (1987) National Wealth and Individual Happiness. In Grunert, K., and Olander, F. (eds.) *Understanding Economic Behavior.* London: Kluwer Academic.

Von Weizacker, E., Lovins, B., and Lovins, H. (1996) *Factor Four: Doubling Wealth—Halving Resource Use: A Report to the Club of Rome.* London: Kogan Page.

Wackernagel, M., and Rees, W. (1996). *Our Ecological Footprint: Reducing Human Impact on the Earth.* Gabriola Island, BC: New Society Publishers.

Warner, K. (2002) Linking Local Sustainability Initiatives with Environmental Justice. *Local Environment.* Vol. 7, No. 1, pp. 35–47.

Warren, B. (2000) Taking Out the Trash: A New Direction for New York City's Waste. OWN and Consumer Policy Institute/Consumers Union, available at http://www.consumersunion.org/other/trash/trash1.htm (accessed February 6, 2004).

Wexler, H. (2000) Bus Riders Union TV documentary. Dir. Wexler and Demetrakas.

Wilkinson, R. (1996) *Unhealthy Societies: The Afflictions of Inequality.* London: Routledge.

Wondollek, J. M., and Yaffee, S. L. (2000) *Making Collaboration Work: Lessons from Innovation in Natural Resource Management.* Washington, DC: Island Press.

World Commission on Environment and Development (1987) *Our Common Future.* Oxford: Oxford University Press.

Yin, R. (1994) *Case Study Research: Design and Methods* (2nd edition). Beverly Hills, CA: Sage Publishing.

Index

About the Author

Julian Agyeman is Assistant Professor in Urban and Environmental Policy and Planning at Tufts University. His research focuses on the nexus between environmental justice and sustainable development at local, national, and international levels. He is the co-editor of *Local Environment: The International Journal of Justice and Sustainability* and *Just Sustainabilities: Development in an Unequal World* (MIT Press, 2003).